Sustainable Energy Consumption and Society

ALLIANCE FOR GLOBAL SUSTAINABILITY BOOKSERIES
SCIENCE AND TECHNOLOGY: TOOLS FOR SUSTAINABLE DEVELOPMENT

VOLUME 7

Aims and Scope of the Series

The aim of this series is to provide timely accounts by authoritative scholars of the results of cutting edge research into emerging barriers to sustainable development, and methodologies and tools to help governments, industry, and civil society overcome them. The work presented in the series will draw mainly on results of the research being carried out in the Alliance for Global Sustainability (AGS).
The level of presentation is for graduate students in natural, social and engineering sciences as well as policy and decision-makers around the world in government, industry and civil society.

Sustainable Energy Consumption and Society

Personal, Technological, or Social Change?

By

David L. Goldblatt
American Association for the Advancement of Science,
Washington DC, U.S.A.

Springer

A C.I.P. Catalogue record for this book is available from the Library of Congress.

ISBN 1-4020-3086-X (HB)
ISBN 1-4020-3096-7 (e-book)

Published by Springer,
P.O. Box 17, 3300 AA Dordrecht, The Netherlands.

Sold and distributed in North, Central and South America
by Springer,
101 Philip Drive, Norwell, MA 02061, U.S.A.

In all other countries, sold and distributed
by Springer,
P.O. Box 322, 3300 AH Dordrecht, The Netherlands.

Printed on acid-free paper

springeronline.com

Printed in the Netherlands.

Contents

List of Figures

List of Tables

List of Abbreviations

BSE	Bovine spongiform encephalopathy ("mad cow" disease)
DSM	Demand-side management
EM	Ecological modernization
ETH	Swiss Federal Institute of Technology
GMO	Genetically modified organism
GNP	Gross national product
LCA	Life cycle assessment
LDC	Less-developed country
NGO	Non-governmental organization
PC	Public communication
RC	Risk communication
SCOT	Social construction of technology
SST	Social shaping of technology
STS	Science-Technology-Society

Preface

Origin of this book

In 1998 the Swiss Federal Institute of Technology (ETH) initiated the 2000 Watt Society, a University-wide research program meant to help Switzerland realize a dramatic reduction in its future energy use. A project within the initiative "Sustainability in the ETH domain," the 2000 Watt per capita Society sought to promote "the gradual introduction of a way of living and working that requires only one-third of current energy consumption but still delivers an improved quality of life."[1] Two thousand watts[2] is roughly the current world average per capita energy consumption; it was the average level in Switzerland in the 1950s; it is some three times less than current Swiss usage;[3] and reducing to that level again (in concert with other industrialized nations) would greatly facilitate stabilization of atmospheric carbon dioxide (CO_2) levels in the long-term, given projections about world population growth and exploitable energy resources [Imboden 1999].

[1] http://www.novatlantis.ch/frames_e.html.

[2] A watt is a power term equivalent to 1 joule/second. Non-engineers generally prefer to think of 2000 watts per capita in terms of energy use per unit time such as joules/second or kilowatt-hours/hour. For Switzerland, 2000 watts per capita is equivalent to a primary annual energy consumption of about 65 gigajoules (65 billion joules) per capita. Although less generally communicative, the 2000 watt name is clean and trim, and since it has stuck we will use it throughout the book. (Primary energy is energy embodied in natural resources that has not (yet) undergone any anthropogenic conversion or transformation and that therefore has not been subject to conversion and distribution losses.)

[3] and, for comparison, about six times less than current per capita energy use in the United States.

That year I had the privilege to participate in a small evaluative project for the 2000 Watt Society under the direction of the professors who would become my principal advisors for this study [Goldblatt 1998]. Shortly before that I had worked on a project on energy and sustainability indicators at the ETH's Energy Analysis Research Group and subsequently Centre for Energy Policy and Economics, focusing specifically on communicative indicators of energy consumption for laypeople. The idea for the study this book describes was born of the confluence of these two earlier projects, and it was also an opportunity to pursue older academic interests of mine, including consumption and the environment and, to some extent, ecological economics. The study thus emerged as an exercise in the theory and practice of sustainable consumption and energy consumption.

Widening the focus to consumption in general (beyond purely energy consumption) followed naturally from the 2000 Watt Society challenge, especially because the 2000 watt per capita target includes both direct and grey (i.e. embodied) energies.[4] It was clear to most involved that achieving 2000 watts per capita in a reasonable time frame would require a range of far-reaching institutional, technological, and social initiatives, and not just an efficiency revolution (perhaps a "sufficiency" revolution as well). The daunting scope of innovation and reform hypothetically needed led quickly to a re-constriction of the practical research to the areas naturally and traditionally strongest in engineering universities: furthering energy-oriented innovations in engineering and science (especially in reducing production and transformation losses), and applying conventional techno-economic tools to energy analysis and policy. In taking up the original challenge, my training, experience, and philosophy lead me to draw a wider circle to consider both science and social science, which turned the inquiry into a sustainable consumption exercise. Appropriately, I conducted this study as both a researcher at the ETH's Centre for Energy Policy and Economics and as a doctoral student in the Department of Environmental Science. Sustainable consumption is perhaps best approached through a multi-disciplinary lens in which sociology figures highly (more on this in Chapter 1). This is in fact the lens I have employed.

The communicative dimension of the study developed, as mentioned, from my previous work on indicators and experience with several software programs for household energy accounting (especially Dürrenberger and

[4] *Grey energy*, also known as *embodied energy* or *energy content*, is the total quantity of energy that is directly or indirectly used for supplying a product or providing a service. It includes the often-overlooked energy involved in generating and transporting raw materials or components earlier in the production chain.

Hartmann's Personal ECO_2 calculator,[5] which we would enhance and deploy in the experimental part of the study). These types of programs had become popular, and have since further gained in popularity, with agencies and NGOs working with the public to stimulate awareness and resource conservation in the home. The Center for a New American Dream's recent Turn the Tide campaign[6] is typical of (American) environmental NGOs' exhortative, bottom-up approach to involving the public in general conservation efforts. By accessing a web-based computer program, the public are shown nine simple steps they can take in their own homes to better the environment: reduce car travel; substitute for beef and shrimp meals; eliminate junk mail; replace standard light bulbs with energy-efficient compact fluorescent bulbs; move the thermostat (down in winter or up in summer); eliminate lawn and garden pesticides; install efficient showerheads and low-flow faucet aerators; and inspire friends. Notably, four of these involve direct energy conservation, and at least another two, indirect, grey energy. The web site calculates the environmental benefits of the individual user's taking these actions, and it also shows the collective impact of the steps group members say they have taken.

Aiming at the householder's direct behavioral change, these software programs are commendable tools as far as they go. (Turn the Tide's incorporation of a collective dimension by tracking the combined effort of its members is especially progressive and crystallizes the attempt to grow social units to a critical mass, at least in cyberspace.) But the energy versions of these sorts of programs for Switzerland, like ECO_2 – which take advantage of energy's convenient property that its use in society is a reasonably comprehensive surrogate for general environmental insult and that therefore energy consumption rates are arguably (environmental) sustainability indicators[7] – showed that virtually no end-user, no matter how conservation-minded or technologically advanced in her household, could presently meet or even approach the 2000 watt standard in her daily life in Switzerland. Grey energies alone usually pushed the user over the limit.

[5] Gregor Dürrenberger and Christoph Hartmann, Der persönliche ECO_2 Rechner, Interview-version 2.0, 2001, ETH Zürich. Henceforth referred to as *ECO_2* or *ECO_2-interview version*. Despite the name, the interview version included energy but not CO_2 accounting, since this would have overburdened lay users with information and was considered unnecessary for the interviews' aims (see Chapter 4).

[6] http://www.newdream.org/turnthetide/

[7] This is notwithstanding the great difficulty in defining sustainability and agreeing on a set of indicators to measure it (discussed further in Chapter 2), and even though carbon intensity is certainly a more accurate measure of the specific climate-forcing potential of specific activities and technologies.

Other factors, constraints in available or useable technology as well as social and cultural forces, were clearly at play. How could one retain the focus on the householder, in both the theory and energy accounting software, but involve the greater universe of factors driving energy use in developed countries? This was the challenge I set for myself in undertaking this research.

This study, then, is partly a response to the conventional laudatory but limited kind of energy/environmental information approaches for the public. It uses perspectives and computer programs characteristic of Turn the Tide as worthy starting points to explore the further limits of information approaches as a lever for changes, on a variety of levels, that support ecological reform. What emerge as an outgrowth are theories, tools, and empirical experiments in energy consumption "monitoring" that provide greater transparency for individual and collective social drivers for unsustainable consumption. Along the way – through literature, theorizing, and interviews – I examine both expert and lay opinions (and sometimes tensions between the two) on the extent of householders' discretionary powers; the relative efficacy of behavioral, technological, social, or economic measures for saving energy; and the extent of different players' roles and responsibility in addressing energy and environmental problems.

This book will show readers something of what I learned in researching these themes and vicariously what participants in the interviews explored in their interactive sessions with the software we developed. The book's multi-disciplinary scope and "trans-disciplinary" treatment should interest academics, students, policy makers, and other professionals working on issues pertaining to energy, the environment, and consumption. As befits its subject matter, the text has also been prepared with the educated lay reader in mind and has been supplemented with numerous summarizing or highlighting text boxes and definitional footnotes. In addition, it is my intention and hope for the book to contribute to a wider discussion of energy consumption and the environment and greater public deliberation on current and alternative approaches.

Environmental perspective

I will be explicit about my environmental perspective and trans-disciplinary aims from the outset. (Although it will be argued that furthering environmentally friendly consumption is not always necessarily, or even primarily, best gone about using explicitly environmental criteria.) In addition, "problematizing consumption as a social activity" [Cohen 2001] or even using terms like *overconsumption* can imply a specific normative stance. This will be spelled out in the first few chapters.

Differences in researchers' values relating to environmental change and quality strongly influence their views of consumption vis-à-vis the environment. Specific valuation of the natural and physical environment, especially as opposed to human capital and the services it provides, determines whether or not environmental "change" is seen as "harm" or "damage" [Stern 1997b]. When judgments are made about the environmental change associated with consumption trends, "different interpretations of sustainability and environmental quality yield different answers to the question of whether past and present household metabolism {and other consumption} rates are in overall compliance with sustainability objectives" [Uiterkamp 1998]. At issue is not whether increasing consumption has negative environmental consequences, but rather whether the environment is being degraded beyond certain biophysical and ecological thresholds and, if so, whether the benefits of the present consumption trajectory outweigh this [Uiterkamp 1998].

At a more basic level, values intrude into the assessment of the reliability of environmental data as well as the legitimacy of environmental standards and targets like the 2000 Watt per capita Society. A collection of facts marshaled in support of one position is easily countered by another collection supporting a contrary view [Tatum 1995]. Although environmental scientists maintain a clear consensus about the severity of worldwide environmental degradation, and adduce ever-greater and more comprehensive bodies of supportive data, these data are usually insufficient to command general agreement in civil society about past and present environmental conditions, let alone likely future conditions, and even less so to dictate policy responses. Whose data should one believe and act on? My reliance on the consensus of the community of scientists on the gravity of environmental problems and the specter of anthropogenic global climate change reflects a normative element in my choice of information sources, albeit an increasingly scientifically compelling one. This study takes the environmental crisis – and the encyclopedic data and studies documenting the environmental impacts of consumer societies – as a starting point.

Structure and content of the book

This book makes extensive use of literature and studies from a range of disciplines. This is to be expected in a problem-oriented, "transdisciplinary" undertaking; such a difficult challenge as sustainable energy consumption requires using whatever help is available wherever it may be found. The reader may be surprised at the lengths of sections devoted to other researchers' theories and findings. This is to derive the maximum benefit from prior and contemporaneous studies in a variety of fields while using them as components in a larger synthesis. The extended discussions of issues, theories, and argumentation in the first three chapters provide valuable input, and culminate logically, in the experimental software development and interviews described in Chapter 4.

Chapter 1 builds the case for sustainable consumption before describing the book's particular information approach to it. It describes the reasons for the environmental challenges posed by Northern[8] consumer societies and consumerist development patterns, and it briefly traces the evolution of the relatively newer policy focus on consumption. The penultimate section theorizes on individuals' and society's degrees of discretion to deviate from their consumption "trajectories," an important concept that will permeate the larger study. The concluding section lays out the book's research aims and general questions and hypotheses.

Chapter 2 completes the characterization of sustainable consumption (or at least *un*sustainable consumption) with an in-depth examination of consumption's driving forces, dynamics, and potential intervention points. Starting with a review of the conventional focus on efficiencies and patterns of energy and resource use, it shows why levels (scale) of use must also be confronted, and with a fuller palette of purviews and approaches from across the social sciences. The chapter critically reviews and synthesizes analyses from a wide range of fields, including psychology, sociology, anthropology, and economics. Chapter 2 thereby provides a full background and expert account of the factors constraining householders in their energy use, which can be compared with lay perspectives voiced by householders themselves in the course of the interviews (Chapter 4).

Chapter 3 focuses on the public's knowledge, understanding, and participation in communication about energy consumption and the larger issues surrounding it. The chapter extends the development of the discretion construct introduced in Chapter 1 and adds another research-related dialectic, Energy-Revealing vs. Social-Revealing. It critically surveys literature on energy "knowledge" and shows how the two theoretical constructs or frameworks are useful both for the analysis of energy consumption and for letting people examine volitional and non-volitional aspects of their use of energy. Section 3 places the proposed extended knowledge-information approach in the context of risk communication

[8] Following convention, industrialized nations are often collectively referred to in this book as the *North* and less developed countries (LDCs) as the *South*. This is notwithstanding Australia and New Zealand's southern location and the fact that nations of the North contain relatively underdeveloped sections, while developing nations may have an upper and increasingly "middle" class whose technologies and consumption lifestyles resemble those of developed nations.

theory and section 4 sets the stage for the software development and interviews of Chapter 4.

Chapter 4 is the study's experimental section. It describes the novel re-configuration of pre-existing energy accounting software, combining household and Swiss national modules to illustrate various conservation and intervention possibilities and other issues from the public's perspective. It reports on the deployment of this software in in-depth (pilot) interview sessions with a cross-segment of laypeople, and it thematically presents the results of a thorough data analysis. Among other things, this empirical work constitutes a pioneering look at several basic issues in the theory and practice of "monitoring" for (householder involvement in) the ecological modernization of consumption.

Note that the populations drawn on most often for examples in the theoretical chapters 2 and 3 are American (US), Swiss, and Dutch. This reflects the quantity, quality, and availability of English language studies from these countries as well as the author's own personal experiences living in each of these cultures. The experiment described in Chapter 4 used Swiss householders (and other primary sources in German, French, and English), since the research and modeling work was done in and for Switzerland. As noted in Chapter 1, however, the theory and communication methods developed are meant to be fully applicable to most Northern industrial societies, since many of the basic driving forces for consumption are increasingly homogenized across the developed world or even globalized.

The concluding **Chapter 5** discusses achievements and implications for theory, research, policy, and future energy information programs and initiatives, either currently being implemented or still hypothetical.

David L. Goldblatt

August 2004

Acknowledgments

I am indebted to my two principal doctoral advisers, Prof. Daniel Spreng and Prof. Dieter Imboden, for so readily welcoming me as an American into Swiss academia, for guiding and encouraging me in my sometimes unconventional ideas about energy, and for helping to fund my stay at the ETH. Dieter Imboden's leadership and projects provided my entrée into the 2000 Watt Society program, and I benefited from both his general management and incisive comments over the course of this research. It was a privilege and pleasure to work under Daniel Spreng in my time at the ETH as I moved from assorted ad-hoc projects at his small research group to full-fledged doctoral work at the Centre for Energy Policy and Economics. Thank you for the innumerable discussions, suggestions, and the trust and latitude you provided me to pursue my brand of transdisciplinary research.

My former colleagues at ETH's Centre for Energy Policy and Economics and sometime office-mates – Lukas Weber, Andrea Scheller, Shonali Pachauri, as well as Bernard Aebischer – gave me good guidance, help, and camaraderie during the years I was there. Thanks to them and others in the sustainability group who offered feedback on earlier versions of Chapter 3 at Lake Thun. In addition, Doris Müller, Igor Bangerter, Martin Jakob, and Marco Semadeni provided important logistical support, and Prof. Eberhard Jochem helped me broaden my ideas and contacts in the field of environmental social science.

Thanks to Novatlantis for financing the majority of the research, and to the Alliance for Global Sustainability for funding part of my preliminary research on energy indicators. Special thanks to AGS' Joanne Kauffman for recognizing the value of the work and helping me through the publication process. Kluwer's Nathalie Jacobs was also of great help.

Doctoral research is at times a strange and solitary endeavor, especially for an expatriate working on a theme as multi-disciplinary and maverick as sustainable consumption. Especially in the beginning, in the course of extensive literature reviews, I enjoyed the "company" of a large number of researchers who have published in the literature, only a smattering of whom I had the privilege to meet over the past several years. To all these authors, too numerous to list here, whose research and writings are so manifestly important to my own work but with whom I never had the opportunity to correspond: all misattributions, misinterpretations, or instances of misuse are entirely my own.

Several people I did meet deserve specific mention. Thanks to Claude Bovay and Urs Dahinden for sharing their time, reports, and primary materials on past Swiss research; Maurie Cohen for instructing me on the intersecting fields of ecological modernization, industrial ecology, and sustainable consumption; Florian Kaiser for his discussions, research, and papers on ecological behavior; Reid Lifset, for enriching my stock of literature and ideas on consumption; John Manoochehri and Laurie Michaelis for keeping me updated on their trailblazing efforts in applied settings and policy circles; Hal Wilhite for his influential work on environmental sociology and anthropology; and Tom Whiston for his instruction and inspiration on public environmental indicators and communication.

Many thanks to Gregor Dürrenberger and Christoph Hartmann, whose original ECO_2 software and collaboration on the interview version were invaluable in actualizing and testing my theoretical notions.

Special thanks to the environmental sociology/policy group at Wageningen University for the opportunity to pay an extended visit and learn and exchange ideas. In particular, Bas van Vliet, Sander van den Burg, and Gert Spaargaren's research and discussions were very helpful in situating my approach within environmental sociology, putting terms to constructs, and connecting to related research in the Netherlands.

Thanks to all of the anonymous interview subjects for their time and effort and to the CEPE staff who braved the early program versions as test subjects.

My family's love and support, in Zurich or long-distance from the US, was indispensable and greatly valued. I thank my mother and father, Brenda and Michael Goldblatt, for their life-long encouragement. Their great hearts, compassion, and tireless work for people and causes continue to inspire me. In addition to offering periodic sage advice, my brother Eric also helped me work through some issues in data analysis, and thanks also to my sister Alisha for her careful reading of an earlier version of one of the chapters and her help with chapter titles. Thanks to Rhoda Schnur for her support and encouragement and most recently for providing the inspirational alpine setting for completing the final substantive revisions.

Paraphrasing P.G. Wodehouse, I dedicate this book to my little daughters Thalia and Livia, without whose never-failing sympathy and encouragement it would have been finished in half the time, and with whom I look forward to laughing together over this quip and much else besides in a healthier world than the present one.

Finally, I'm immensely grateful to my wife Aviva, without whose loving encouragement, tolerance for my late nights, and unstinting devotion to the children this book would (seriously) never have seen the light of day, and without whose careful proofreading and political scientist's eye it might not have deserved to.

Chapter 1

SUSTAINABLE CONSUMPTION AND THE PUBLIC'S ROOM TO MANEUVER IN ENERGY USE

1. INTRODUCTION

Old themes are resurfacing in recent discussions of the energy problem. The energy bogeys of the 1970s, once thought banished by a new world order and post-industrial information societies, are back to haunt us. In the United States, soaring gasoline prices, concern over supply security, and acid rain and mercury from coal-fired power plants are all prominent in the news. The attacks of September 11, Al-Qaeda, and the war in Iraq have generated some renewed focus on the continued vulnerability of the West to geopolitical instability in the Middle East, and in the wake of the 2004 oil price shocks, financial markets are reacting anxiously to the possibility that once again high energy prices could fuel general price inflation or derail the economic recovery. The next record-breaking flood, fire season, storm, or heat wave – perhaps with the help of a Hollywood disaster movie – will also generate some discussion of the newer breed of environmental problems associated with climate change, and a few of the better informed will make the connection with CO_2 and energy use. In longer discussions, American experts may mention Alaskan oil drilling or renewable energy, depending on their political orientations, as a solution for several of these interconnected problems. A few seasoned observers reflect wistfully on the glut mentality of the '80s and '90s auto drivers and lawmakers who so thoroughly forgot the lessons of the 1970s. Sustaining these higher prices with gasoline taxes, if necessary – the standard textbook remedy but anathema in the best of times – is out of the question, so the introduction of the hybrid sports utility vehicle is anticipated as the next-best solution.

In these discussions, certain older topics and terms tend to be conspicuously absent. Radical *energy efficiency* advancements across most sectors of the economy, technically feasible but economically or politically difficult, are at best mentioned only as a far-off possibility. *Energy conservation*, sullied by association with cutbacks and self-denial in the 1970s, is passé. World energy *consumption*'s implication in two of the greatest challenges to come in this century, according to the preponderance of scientific opinion – further accelerated global environmental degradation, in particular the consequences of climate change, and the probable re-concentration of stagnating world oil production in the Middle East – is not really confronted. The question seems to be not when but if we will begin actively to prepare for a transition to a post-carbon energy basis for the world economy. Nobody in power or prominence openly questions the long-term economic and environmental feasibility of increasing or maintaining Northern levels of energy and resource consumption while simultaneously encouraging the South's, especially Asia's, headlong race to match them.[9] In short, the *sustainability* of societies' resource and energy systems, and that of the economic and ecological orders whose fate is bound up with them, is seemingly taken for granted. And if environmental *sustainability*, once at least paid rhetorical lip service, lacks political currency, *sustainable consumption* is unmentioned or unmentionable. But sustainable consumption is a key part of the solution.

This book looks at energy as a central question of sustainability, and sustainability as largely a matter of consumption. Its objective is stimulating a greater, more effective involvement of the central actors, the consumers themselves, specifically through the vehicle of improved information means, by showing them how and where they can "make a difference."

This book is not an anti-consumerist polemic. In setting the background, the first two chapters do, however, focus critically on general consumption, consumerism, and consumer societies in the developed world. In this way energy consumption, which is more often treated independent of other types

[9] No fault is found with attempts to lift population (subgroups) out of energy poverty. It would also be less problematic were the Southern economies actually systematically to leapfrog to first world levels of energy services using radically energy and resource-sparing technologies.

of consumption,[10] takes its place in the broad spectrum of consumption types. This is appropriate since energy use responds to many of the same general driving forces, is shaped by many of the same social practices and norms, and can be viewed through some of the same lenses that analysts use to study general material or service consumption. The obverse approach, treating material consumption as indirect, embodied energy, is also useful and plays a large role in the energy model and interface developed for the field study (Chapter 4).

Having drawn a wide circle of world problems in the opening paragraphs, we will quickly bring the focus down to environmental degradation, and especially climate change. This book is specifically concerned with how consumer societies, technologies, consumerist activities, and their driving forces might be influenced to ameliorate negative environmental consequences of their attendant use of energy and other natural resources. Individually, psychological and ethical arguments make perhaps the strongest case against excessive consumption or consumerism. Lowering consumption may reduce the pressures of competing in the modern economy and may ward off excessive individualism and materialism. On a collective level, social, environmental, and, indirectly, national security factors stand out as the most compelling reasons for curbing excessive consumption [Goodwin 1997]. Again, however, in this study, psychological, social, and economic problems stemming from consumption will only be addressed insofar as policies directed at them (or arguments concerning them) may be effective in reducing or altering environmentally significant consumption and alleviating its negative environmental consequences. The environmental problematique surrounding consumption, and not the array of other potentially related problems, is of central interest here. Similarly, the scope will be limited to consumers in developed countries with their familiar capitalist democracies and relatively open news media, even though the rapid pace of Southern industrialization and the emergence of "new consumers"[11] there add an urgency to address consumption both there and in the North.[12]

[10] The designated experts, researchers in the field of energy analysis born in the wake of the 1970s energy crises, tend to have little or no contact with analysts and academics studying general consumption in many of the social sciences.

[11] See [Myers 2004].

[12] We reject the argument raised in the failed Kyoto Protocol negotiations over limiting CO_2 emissions that the United States need not reduce its emissions until the less developed countries commit to similar reductions. The first industrial powers have obligations and not just privileges in a globally industrialized world. They have great economic

2. THE CONSUMER SOCIETY

"Perhaps more than any other activity, consumerism has become the lynchpin of modern {Northern} economic, political, social, and personal life" [Schor 1995]. Yet historically, this is a relatively recent development.

Enormous increases in production and consumption over the last century completely transformed Northern societies. World industrial output is now more than fifty times greater than in 1890, most of the expansion having occurred in the North after 1950 [Ponting 1991, Pfister 1996]. There, fossil fuel use permitted a huge increase in material and energy throughput and labor productivity. At the same time, environmental and social costs of fossil fuels were externalized, permitting material and energy use efficiencies to be largely ignored. If their economies had internalized these costs, they would have emphasized improved resource productivity, which might have meant a lower material standard of living but fewer environmental problems [Røpke 1999], [Ponting 1991]. After fossil fuel use the second major enabling factor for this growth was an enormous transfer of additional natural resources from the South. Division of labor, urbanization, competition, and other features of capitalist industrialization are also mentioned as enabling factors for the modern growth of consumption [Røpke 1999].

The socio-economic phenomenon of consumerism is associated with the 20th century, especially the post-World War II era. The "consumer society" has risen in tandem with the development of mass production, the concomitant decline in home production, and the increased consumption of ready-made and time-saving products [Gatersleben 1997]. Consumption has been enriched by "the increased variety and many-sidedness of the consumption basket resulting from the higher amount of discretionary spending and leisure" ([Uusitalo 1982] as quoted in [Gatersleben 1997]).

Consumerism presupposes large opportunities for economic consumption – availability of goods, materials and services; accessibility; relevant, available information; and a stable price system – all of which have increasingly obtained in Northern economies. Prices have continually decreased relative to income, and opportunities to purchase on credit have been widely introduced [Gatersleben 1997].

The social changes facilitating the consumer society, including the decline of community and elevation and isolation of the individual [Lintott 1998], stand out even more markedly than the economic changes. Like other social systems, the consumer society has required the service, or creation, of

---

opportunities besides in the new post-carbon energy industries that will emerge, if they would only seize them.

cultural attitudes to support itself. Social pressures, advertising, shopping culture and infrastructure, and government economic policies all work to perpetuate consumerism [Durning 1991].

From the perspective of social psychology, the consumer society makes the possession and use of an increasing number and variety of goods and services a main aspiration and source of happiness, status, and success. In particular, it links self-respect to one's level of consumption as compared with others ("positional consumption") [Goodwin 1997b]. Consumerism has become deeply rooted in the psycho-cultural fabric of life in the developed countries, most obviously in the US, but increasingly in other developed (and, where possible, less developed) countries as well. Advertising and marketing encourage wants, and through consumption people also attempt to satisfy non-material wants such as identity-forming.

3. CONSUMPTION TRAJECTORIES IN WESTERN EUROPE

Although the United States is known as the king of consumerism and the largest world consumer of resources and energy, this section focuses on Western Europe to give an indication of the scope of the problem across the developed world. More than fifty years after World War II, Western European households'[13] use of resources and energy in many cases does not yet show signs of saturation and stabilization at long-term sustainable levels.[14] In post-war Western Europe, social and demographic phenomena, including increases in the number of households and a simultaneous decrease in their size, population growth, rising incomes, and expanding infrastructure, consistently offset technological efficiency gains [van der Wal 1998]. These phenomena reflect significant social transformations in values and lifestyles. Self-fulfillment, personal development, independence, and other hallmarks of individualism have increased in social value. Western European religious and political institutions have correspondingly decreased in importance [van Diepen 1998].

[13] Tellingly, most Western European countries have removed the notion of family and kinship from their official statistical definition of *household* and instead define it operationally as a unit of people who share some "moments or attributes of consumption of some kind" [van Diepen 1998].

[14] This seems too long to constitute an early inefficient phase in the transition towards optimal patterns of resource use (squandering as an "inevitable feature of evolutionary processes" [Uiterkamp 1998]).

In recent decades the rate of growth in the number of households has tended to exceed the population growth rate. These households have demanded more goods and services of the infrastructure [Noorman 1998b]. Between 1960 and 1990, the population of Europe (excluding the Soviet Union) increased by 18%, from about 430 million to over 500 million. At the same time, the number of households increased at double the rate, implying a general shift from larger to smaller households. Households of five or greater have been steadily declining. The proportion of one-person households increased the most in Sweden, the Netherlands, and Switzerland. In the last, this proportion had increased to 32% by 1991. In general, the largest decreases in household size, changes of more than 20% over 30 years, occurred in Scandinavia, Switzerland, and the Netherlands [van Diepen 1998]. This household "dilution" impacts resource use not only from the increase in the number of separate dwellings but also from economy-of-scale effects. In energy and electricity use, household scale effects are common: larger households are relatively more energy efficient and produce less waste per member than smaller households [van der Wal 1998].

As for economic trends connected to resource use, real prices for energy have generally fallen over the years since the oil price shocks of the 1970s[15] and are likely to have stimulated demand. Householders have acquired more appliances and used more services even as energy used per appliance or service unit has declined. Automobile use has increased steadily, and both the average number of autos per household and their rates of private use have increased. Real auto fuel prices fluctuated but did not markedly increase over the last decades of the 20th century despite the two oil crises in the 1970s [Linderhof 1998]. These price trends represent part of the increasing opportunities and abilities that have brought about increasing household consumption.

The general conclusion of a large Dutch study on the flows of energy and resources through households in the Netherlands, sometimes referred to collectively as *household metabolism,* is that the rates of these flows have increased rapidly since the 1950s, and that considering the driving forces, Dutch householders are continuing on an unambiguously non-sustainable path, at least in the short-term [Uiterkamp 1998]. While demand for residential heating has dropped back to the level of the 1960s, electricity consumption per household has risen mainly because of increasing numbers

[15] Outside of a few transient price shocks like the one during the first Gulf War in 1990-91. The large oil price increases that have occurred during the present war in Iraq may not be transient, as they are being driven by surges in world (Asian) demand as well as geopolitical factors.

of appliances. Penetration rates of refrigerators, washing machines, and televisions increased from almost zero to over 100% [van der Wal 1998]. The "enrichment of Dutch consumption patterns" is marked by an increased use of goods related to entertainment and personal development over the last twenty years. Another factor is increased time and money spent on vacation, as well as increased per capita private car ownership and distance driven [Gatersleben 1998].

A parallel study of the consumption habits and outlooks of households in the United Kingdom yielded data on the phenomenal expansion of consumer purchases there from 1954 to 1994. The overall rate of increase in purchases during that period was 100%, while the single largest increase for a category was recreation and entertainment at 400%. The acquisition of goods within the sub-category of durable entertainment goods like stereos and televisions increased by 3500%! Corresponding rates for general domestic appliances, communication, and travel were 385%, 341%, and 293%, respectively [Jackson 1999].

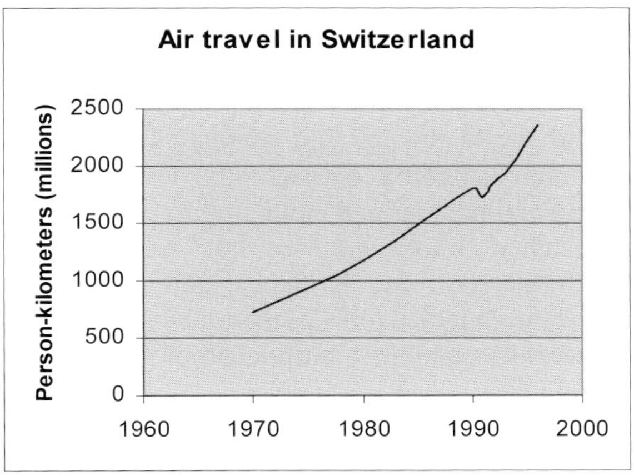

Figure 1-1. Air travel in Switzerland. Source: Swiss Federal Statistical Office, 2002.

In Switzerland, per capita heated floor area for residential housing doubled between 1960 and 1997, even while technical improvements steadily reduced the unit energy consumption to 500 megajoules per square meter. Apart from heating, per capita electricity consumption tripled between 1960 and 1997, despite efficiency improvements of 20-25% since 1973, because of household dilution, greater use of established services, and diffusion of appliances and new services [Aebischer 1998]. For example, an estimated 11 times more electricity was used for lighting in the household sector in 1990 than in 1950 [Schwarz 1996]. For another example, the percentage of Swiss households equipped with at least one computer more than tripled between 1990 and 1998 (to 51%), while 7% owned two or more [Swiss Federal Statistical Office 2002]. In the transportation sector, average personal travel (excluding air) increased from 9400 km per capita in 1970 to 14,300 km in 1990. Automobiles consume some 60% of the energy used in the transport sector in Switzerland, and their number now exceeds the number of households. Car ownership is highly unevenly distributed: a quarter of households do not own a car while 30% own more than one. An average of 50% of trips by car are for leisure, and leisure time has gone up steadily [Aebischer 1998]. Air travel has also risen sharply although unevenly across the population (see Figure 1-1).

4. NORTH-SOUTH DISPARITIES, NORTHERN CONSUMERIST MODELS, AND GLOBAL RESOURCE CONSTRAINTS

To place the developed countries' consumers in a global context, the majority of them are members of the global "affluent" class. As of the early 1990s, this group constituted approximately one-fifth of world population but earned 85% of world income and accounted for over 80% of global resource consumption, according to gross estimates. As noted, their numbers are growing with rising incomes and emulation by segments in the LDCs. As of 1992, they consumed on average three times as much water, ten times as much energy, and nineteen times as much aluminum as the average of the other two classes [Durning 1991]. As of the mid-1990s, they generated more than 75% of the world's municipal and industrial pollution, a large majority of hazardous and nuclear waste, and had been responsible for about 80% of the cumulative anthropogenic CO_2 emissions since 1950 [Symposium Sustainable Consumption 1994]. The "global middle class," which comprised three-fifths of the world population in the early '90s, had historically consumed relatively moderately: Durning held them as a model for the rest. The remaining fifth lived in destitution [Durning 1991]. After a decade of further globalization, Conca contends that the "sustainable middle," especially in the Southern emerging markets, is being "squeezed" on either end, globalization drawing some into the ranks of high consumers while marginalizing and impoverishing the rest [Conca 2001] [Princen 2002]. These enormous disparities and inequities in wealth distribution, resource consumption, and waste generation have been the subject of frequent discussion and contention between the North and South in international environment and development forums.

For many resources, a line cannot easily be drawn between developed and less developed countries. Complex production-consumption chains wind through both worlds, livelihoods are linked, ecosystem damage occurs in both, and both developed and less developed countries can bear the resulting economic and social impacts. Consumption of three indicator renewable resources – cereals/meat, wood fiber, and fish – is rising everywhere. Demand is not just from "lifestyle" demands of the wealthy Northern countries, but also from people's basic needs such as housing and nutrition [Matthews 1999]. Yet for those resources for which there is no easy substitute, there is no doubt that high levels of resource consumption in the North leave less (or already, as in the case of many fish stocks, virtually none) for people of less developed countries. Industrialized nations' huge, disproportionate draw on stocks of resources brings them near their limits or system thresholds faster, sometimes necessarily to the exclusion of other current or future societies. Destabilization of the atmosphere through disproportionately high greenhouse gas emissions can be viewed similarly. The effect of consumption patterns and lifestyles in the North on future consumption and production potentials in the South was a commonly raised concern at the 1992 UN Conference on Environment and Development (the Earth Summit) in Río de Janeiro [Measuring Changes 1998]. Distribution schemes for apportioning countries' or people's relative share in global resources more equitably are not discussed here.[16]

The philosophy of the global economic system (the "neo-liberal regime") does not fully recognize inherent limits on productive or absorptive capacities of renewable resource stocks and ecosystems, or it trusts technology to overcome them. The debate on the means to reduce the North-South disparity – reducing Northern consumption levels (while allowing enough growth in the South to eliminate widespread poverty) or attempting to bring Southern levels up to those of the North – illustrates a basic philosophical and normative divide. Many environmentalists argue that the manifest limits on certain vital resources, limits that have not been or cannot be overcome, make it futile and dangerous to embrace the second path. This side insists that affluent countries leave enough for Southern populations to meet their needs, even at basic levels (for example, see [Haavelmo 1992]). The projections of continued growth in world population in the near to mid-term strengthen their convictions [Lintott 1998]. While both sides agree in theory on the necessity for vast improvements in the methods and efficiency

[16] For example, uniform per capita apportioning of rights to natural resources is the most commonly chosen distribution in ecological footprint and rucksack analysis [Wackernagel 1997], [Schmidt-Bleek 1999].

of resource utilization, the world's power brokers are committed to universal growth as the solution to Southern poverty and inequity. This growth paradigm, embodied in the national political goals of increasing GNP and personal expenditure, predominates as the model for development in the North and most of the South.

This paradigm has seen its ultimate success in the large-scale emulation of or aspiration to high consumption lifestyles by lower social strata in both the developed and less developed worlds. The modern consumption practices and habits of the educated global urban middle class (Durning's "affluent class") have become the example for lower social strata to imitate within Western countries – and, promoted through aggressive advertising on television, the Internet, and other communications media – increasingly in less developed countries.

The global emulation of Western lifestyles is alarming for its environmental consequences and implications for the future. Constituting a form of "anti-environmental modernization" ([Garcia 1998] the opposite of that predicted by the environmental Kutznets curve[17]), it is already credited with accelerating environmental degradation. Were the current US per capita consumption of non-renewable resources, for example, to be imitated by the world population today, total consumption would be on average at least seven times as large as presently [Daly as cited in Honkasalo 1998]. Until the 1970s, average European and Japanese lifestyles were significantly less material- and energy-intensive than the United States', and much less than they are at present. Goldemberg claims that those pre-1970s lifestyles could have been sustainably imitated by less developed countries [Goldemberg 1987]. Current Western European lifestyles do not bear global imitation. The attempt is not only greatly accelerating the ecological decline, but the drain on energy and resources is beginning to be registered by large price increases for raw materials and fees for their transportation and delivery. The oil price shock in the first half of 2004 is being openly attributed in no small part to the exponential growth in the Chinese economy and the concomitant exploding Asian demand for resources. And this engine of demand is just starting to accelerate, as consumer aspirations and lifestyle emulation are actively encouraged by multinational corporate marketers and advertisers. No one seems to have a satisfactory answer to the challenge posed by the environmental consequences of a global proliferation and high-level

[17] The environmental Kutznets curve predicts that at some point after a developing country's GNP exceeds a certain threshold, its environmental quality starts to increase, presumably as people there becomes rich enough to worry about environmental quality.

homogenization of heretofore Northern consumption. The traditional response has been to deny there is a problem.

5. THE EVOLVING INTERNATIONAL DISCOURSE ON SUSTAINABLE CONSUMPTION

Until recently the North's ability to set the agenda in international forums and the South's fragmentation and political ineffectiveness limited official attribution of the causes of global environmental degradation largely to population growth in less developed countries. (This account is based on [Cohen 2001].) By the time of the 1992 Earth Summit, LDCs and allied NGOs were empowered to force a widening of the problem definition to include Northern consumption as a major culprit. Northern countries' relative success in controlling production-related, localized pollution in the 1970s and 1980s, and the emergence of much more difficult global environmental issues like global warming and diminishing biodiversity, also contributed to resetting the international environmental agenda.

Agenda 21, the UN's environmental blueprint that came out of the Earth Summit, clearly identified patterns of consumption and production, especially in industrialized countries, as a chief cause of global environmental degradation. The gravest future danger, it asserted, was the emulation by less developed countries of the developed world's environmentally damaging lifestyles. Although Agenda 21 made some tentative steps towards a consumption orientation to the global environmental problematique, political hamstrings and disciplinary blinders prevented it from recommending much more than the use of improved information campaigns and market price mechanisms.

Thus, politically speaking sustainable consumption is in its origins essentially a Southern concern or at least the product of Southern assertiveness. Its unpopularity and relatively marginal presence on Northern national agendas reflects a continued resistance to a politically threatening issue.

The 1997 joint statement by the Royal Society of London and the US National Academy of Sciences called Towards Sustainable Consumption was a scientific watershed [Royal Society 1997, Cohen 1997]. For the first time, the premier scientific organizations of the US and UK explicitly departed from the decades-old line that population growth in the less developed countries was the (exclusive or) principal root problem in environmental decline and implicated Northern consumption patterns and their associated consumerist lifestyles and values. Subsequently, however, the great pressure brought to bear against American scientists maintaining politically unsavory positions on controversial scientific issues ("speaking truth to power") [see Yankelovich 2004] seems to have largely quashed further progress in advancing the scientific dialog on sustainable consumption in the US.

6. CONSUMPTION OR PRODUCTION: WHICH IS THE BETTER FOCUS FOR ADDRESSING ENVIRONMENTAL PROBLEMS?

An oft-debated issue in environmental policy in industrialized countries has involved the question of to what extent consumption is actually responsible for the environmental predicament, rather than production dynamics and processes and their related economic, institutional, or technological features. Where should the emphasis be placed and the remedial effort applied, on the consumer or the producer side?

Studies of the environmental impacts of consumption typically use individuals or households as their unit of analysis. Households consume a complex and changing mix of goods and services, yet different households' general lifestyles have differential environmental effects. Individuals in households exert a large influence on the type and quantity of goods and services produced, and, through the pressure of their collective demand, they may affect the environmental characteristics of the manufacturing processes [Noorman 1998b].

A general criticism of the household emphasis is that governments', corporations', and organizations' (e.g. the military's) decisions have much greater environmental ramifications than individual householders' choices. The assumption that most consumption in industrialized countries stems directly from individuals and households may be incorrect in many cases. In many countries in the North, corporate and government activity is directly responsible for the majority of energy consumption, pollution, and other environmental damage. The acquisition and use of products and services by industry, government, the military, and other large organizations often constitute a large proportion of GNP (Of course, this demand can be accommodated in a wider definition of consumption. And arguments can be made, on the basis of economic if not political theory, that household consumer behavior indirectly drives a large part of corporate and governmental decisions in Western industrialized nations) [Stern 1997b].

The next level of criticism of the individual or household approach is that it neglects the control that businesses, government, and organizations have over decisions that more directly affect energy and material consumption in the economy, such as fostering certain technological paradigms, setting official and de facto efficiency standards, offering product lines, and using specific industrial production methods [Stern 1997b]. Matthews and Hammond emphasize that poor production methods along with exponential increases in demand are at the root of the liquidation of renewable resources (while at the same time they stress that rising consumption demand for basic needs from less developed countries, and not only resource-intensive lifestyles in the developed world, is driving the bulk of this demand) [Matthews 1999].

The focus on production and in particular producers' power over consumers has older precedents in both policy and theory. Schnaiberg (1980) held that consumer demand in industrialized countries had been determined primarily by producers; and ameliorating environmental damage must focus on producers, production and the politics supporting private business interests. According to Schnaiberg's "treadmill of production," high levels of consumption are necessary for producers' profits, so business and government use various means to keep consumption levels high. Consumer choice is limited to producer offerings, once a lifestyle is chosen or set. The power that business and political interest groups exert on government, especially prominent in the United States, has dictated infrastructural decisions that have "locked-in" many aspects of consumption. For instance, in many areas in the US, public goods like transportation and related infrastructure have been limited, hobbled, or eliminated by the influence of auto manufacturers and oil and trucking companies who have maintained their subsidies and locked consumers into using car use and its attendant lifestyle [Schnaiberg 1980] [Ponting 1991].

For their part, consumer-side interventionists emphasize that the past several decades of enormous growth in resource demand and consumption by Northern households show that there is still a large area of "system-discretionary," producer-independent, environmentally significant household behavior that can be influenced at the level of the consumer. Schnaiberg does not make a good case that changes in production practices and institutions alone are still sufficient to halt the environmental damage from resource consumption, especially given the continued enormous growth in consumer demand. Nor could he claim that Northern consumption rates for many resources could continue to be satisfied along with basic needs in the South, even with such better practices.

In policy circles, demand-side measures to reduce consumption are now commonly considered in addition to technological or regulatory means for transportation, water, and waste management. Driving this emphasis is the perception that utilization growth has overwhelmed unit efficiency improvements in many cases, and that wastes associated with consumption

Definitions of consumption and environmentally significant consumption

Economists' treatment of the environmental impacts of consumption traditionally extends to pollution from various production processes and to some degree to the direct and indirect impacts of economic activity on natural resource stocks and ecosystem productivity. Conventional economics emphasizes the value added from technology, whereas ecological economics emphasizes the importance of that which is added to, i.e. natural, low-entropy resources. This latter quantity is "consumed" in human consumption [Daly 1996a]. From a physics perspective, all consumption consists of entropy-increasing transformations of matter and energy. This entropy increase may be environmentally significant because it degrades the quality of the stock or ecosystem in which the transformation occurs [Stern 1997b].[18]

If consumption is human transformation of materials and energy, then environmentally significant consumption is related to a transformation that "makes materials or energy {stocks} less available for future use, moves dynamically stable biophysical systems toward a different state or, through its effects on those systems, threatens human health, welfare[19], or other things people value" [Stern 1997b]. Such a definition emphasizes the biophysical effects of human social and economic activity in general, and it makes clear that although consumption lies at the "interface of social and natural sciences," appropriate units for environmentally significant consumption are "physical and biological, not economic or social" [Stern 1997b].

Broadly defined, environmentally significant consumption can be a form of ecologically damaging misconsumption on the individual level,[20] harming single groups of resources and/or people who depend on the resources, or it can be overconsumption on an aggregate level. "Overconsumption is that level or quality of consumption that undermines a species' own life-support system and for which individuals and collectivities have choices in their consuming patterns"[21] [Princen 1999, 2001]. (Aggregate consumption concepts are examined further in Chapter 2, section 3.4.3, "Restraint and sufficiency.")

In social science settings, economic units are used for consumption as a matter of convenience and familiarity. We follow this usage most often in this book, where final or end-use consumption is discussed as a surrogate for biophysical transformations. In an economics context, the concept of final consumption assumes a clear distinction from production, whereas it is also possible to view human consumption as an input to labor, in which case there is no such thing as final consumption. However, this is appropriate mainly at the level of fulfillment of basic needs and not at the much higher levels of consumption found in developed countries [Røpke 1999]. Similarly, household resource and energy flows are often measured with economic and energy units, while pointing to the direct and indirect physical and biophysical consumption of natural resources [Noorman 1998b].

[18] The entropy-related understanding of consumption is characteristic of ecological economics, in particular the work of Georgescu-Roegen [Georgescu-Roegen 1971].

[19] *Welfare*, as it is used by the author and others cited in this book, refers to well-being in the economic sense (conventionally assumed to be correlated with economic standard-of-living or spending ability).

[20] The misconsumption may also be physically, psychologically, or economically damaging to the one doing the consuming.

[21] Collective overconsumption is also an ethical issue "because it inheres only in those populations or species that can reflect on their collective existence," and it can become a political problem when "the trends are toward collapse, power differences influence impacts, and those impacts generate conflict" [Princen 1999, 2001].

now often exceed waste emissions from manufacturing processes [Measuring Changes 1998]. The joint consideration of production and consumption patterns has been embraced by many analysts and represents a consensus focus of the UN's workshops on consumption and production. Reportedly, such an approach permits examination of both economic and behavioral social policies, spotlights the full-use lifecycle, and encourages examination of the distributional effects of these patterns [Measuring Changes 1998]. But as the discussion above of Agenda 21's limitations suggests, the recent focus on the demand side is a far cry from embracing sustainable consumption. The economic and technological interventions commonly suggested are instrumentalist and mechanistic, and policy-makers who over-rely on them tend to have a "levers, knobs, and dials" view of the world [Cohen 2001].

Common analytical uses of consumption, such as one which identifies consumption with overall economic activity, are often actually disguised production approaches, or they conflate consumption with other concepts like materialism, maldistribution, population, or technology, thereby losing or obscuring the ecological aspects or driving forces [Princen 1999]. Political theorist Princen calls for a consumption focus, but one defined in an ecological economics manner that permits concentrating on non-purchase, non-commercial [cf. Cogoy 1995], or non-material responses to needs; material provisioning; and material decision chains [Princen 1999]. However, treating environmental problems from such a consumption perspective can generate great analytical and policy resistance: "The reason is that to talk about consumption levels and consumption patterns is to talk 'out of paradigm.' It is to eschew the production perspective and to raise analytical questions that conventional analytic tools – price determination, cost-benefit analysis, even life cycle analysis – cannot comfortably address. It is, ultimately, to raise question{s} of purpose." [Princen 1999].

Sustainable consumption is thus one of the most important but politically difficult of the key issues in the sustainability policy debate. It has been slow to be taken up by researchers and policy makers, not least since in its maximalist forms it can lead to questioning assumptions about wants and needs, the consumerist basis of Western economies, or even the economic growth imperative. Expanding the field and defining the issues so broadly, and considering such a wide range of factors and social institutions as relevant, is to invite criticism, dissent, and controversy.

7. ROOM TO MANEUVER IN ENERGY USE

God, give us grace to accept with serenity the things that cannot be changed, courage to change the things which should be changed, and the wisdom to distinguish the one from the other.

Reinhold Niebuhr, The Serenity Prayer

Sustainable consumption is both politically and practically challenging. Changing household consumption is fraught with difficulty whether it is approached at the individual or the sectoral level. In industrialized countries much of householders' resource and energy consumption is resistant to change. Individuals cannot change their domestic routines and behaviors, or to the extent they can in the numbers who are willing, it would do relatively little to reduce associated draws on energy or resources. This has largely to do with rigid constraints that technology, culture, and institutions impose on broad-scale patterns of consumption. Technologies typically become entrenched or "locked-in" and thereby limit consumer choice within internally propelled systems of commodities, infrastructure, social practices, and institutions and within technological trajectories [Røpke 1999]. In general, the prevailing socio-economic-technical framework tends to limit individuals' capacity to choose and chart their consumption.

In the 1980s, marketing researchers R.R. and N. Dholakia presented a macro-micro model of energy consumption behavior as a series of nested and interlocking choices, in which "macro choices delimit and define the scope of micro choices." "Household energy use is ... not just ... the result of a choice among behavioral alternatives but ... the production of these alternatives is also viewed as the result of a social choice process. In other words, energy use and energy conservation behaviors must be seen within the context of a broader consumption pattern which is socially determined" ([Dholakia 1983], emphasis in the original). The Dholakias used the term discretionary[22] to characterize the individual's scope in making micro choices; the implied term for the constraining macro consumption pattern, as it relates to the individual, would be non-discretionary or less discretionary.

[22] *Discretion*: 1) The quality of being discreet; circumspection. 2) Ability or power to decide responsibly. 3) Freedom to act or judge on one's own (*The American Heritage Dictionary of the English Language*, Third Edition, 1996, Houghton Mifflin). Another version of 3) is freedom to act according to one's own judgment; unrestrained exercise of choice or will (*Webster's Revised Unabridged Dictionary*, 1998, MICRA). All three meanings are relevant to this usage of the term *discretion*. The third meaning is the one most directly intended. Yet the second, the ability to make responsible choices, is closely related. We also hope to foster a societally "discreet" use of energy and resources.

Many factors – economic, technological, social, and others – shape this non-discretionary consumption. Technological and economic factors include technological innovations (for example, microchips for information technology and communications), their consumer applications and diffusion (PCs, cellular phones, and embedded chips in consumer appliances); manufacturing and efficiency standards; energy prices and government taxes and subsidies; corporate locks on technologies and efficiencies (e.g. Microsoft and the Big Three American automakers); and general market imperatives like those for innovation and planned obsolescence. Other factors are more social or cultural in nature, for example; ever higher standards of "comfort, cleanliness and convenience" in the home [Spaargaren 2000b]; and the trends in both developed and less developed countries towards individual and private, rather than collective and public, forms of motorized transportation. Many factors are multi-faceted, and some are deeply ingrained, like the systemic bias towards satisfaction of consumers' needs through commercial markets rather than by non-purchase means. A number of these factors are explored in detail in Chapter 2.

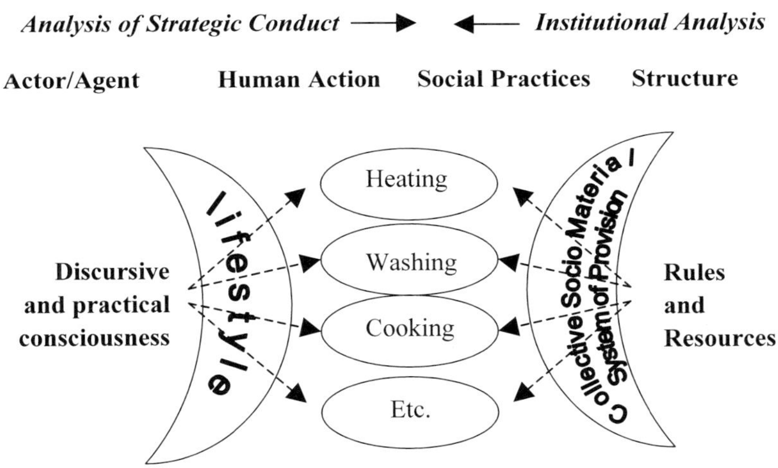

Figure 1-2: An Actor-Structure Model of Consumption [van Vliet 2002]

In sociology, Gidden's structuration theory seems to come closest to this explanatory framework [Giddens 1984]. When knowledgeable and capable individuals act in a social context, they draw on a virtual set of social rules and resources (structure), but in doing so they instantaneously reproduce (perpetuate) these rules and resources. Social (and socio-technical) structures are both enabling and constraining of individuals [Spaargaren 2000a].

Attempts to pursue an "ecological modernization of household consumption" have to recognize this mutual actor-structure interdependence, as is depicted in Spaargaren and van Vliet's conceptual model (Fig. 1-2) [van Vliet 2002]. Here, environmentally-minded individuals depend on systems of provision to make the necessary technological and institutional innovations available, while innovative companies and agencies need a degree of consumer cooperation and acceptance [Spaargaren 2000b].

Thus, some part of a person's domestic energy use is more highly discretionary, and this part may sometimes be measured through economic or psychological devices. A *collective's* or society's "room to maneuver," though (its "degrees of freedom"), is really only discernible through empirical historical study [Spaargaren 2000a].

7.1 Northern society's discretion in energy use

As a thought experiment, we can imagine an energy "discretion continuum" or range for society based on hypothesized requisites for sustainable consumption (more in Chapter 2), using what it has achieved thus far as a measure of the minimum the collective deems currently possible or discretionary. We know that the current potential for end-use energy efficiency[23] development, for example, far exceeds what prevails in the marketplace, and we can hope that the potential for social and economic development also exceeds past and current achievements. Figure 1-3 portrays Northern society's current discretion continuum vis-à-vis resource and energy use, with examples of institutions arrayed on both sides of the discretionary/non-discretionary threshold (those above the border represent theoretical institutional or technological innovations it does not or cannot yet achieve). The vertical lines to the right sketch possible directions for future trajectories.

7.2 Individuals' discretion

Research and policy concerned with household energy use often concentrates on the evidently discretionary end of the range for *individual* actors. Acquisition and use of relatively energy-efficient appliances is to

[23] End-use energy efficiency: the efficiency with which energy is consumed by end-users to produce energy services in such areas as buildings, transportation, and industry.

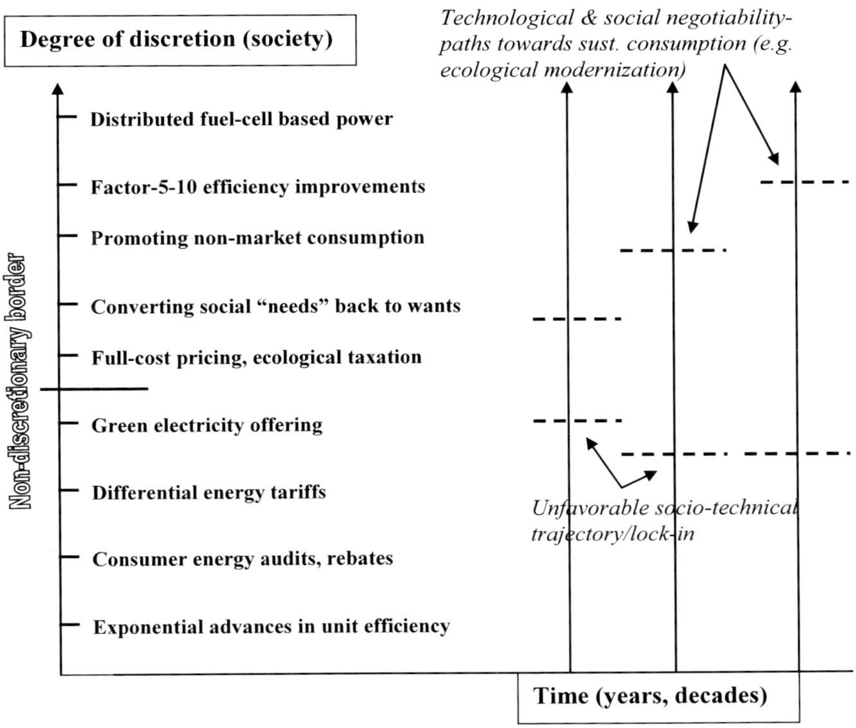

Figure 1-3. A depiction of society's current "discretion continuum" for resource or energy consumption, with future alternative trajectories

some extent open to choice and has often been the focus of direct policies meant to reduce household energy use. Daily domestic practices are to a large extent routines, which can – with the right impetus – be actively de- and re-routinized in less consumptive, more environmentally friendly ways [Spaargaren 2000a]. To a lesser extent, selection of one's "lifestyle" – particularly during early adult years when defining career, social group, and family decisions are made – is amenable to individual influence. In energy studies, a "lifestyle" factor has been theorized to account for empirical variations in energy consumption within groups of similar income, culture, and other attributes [Vringer 1995a], [Wilting 1998] (see Chapter 2, section 3.3, subsection "Culture, Society, and Lifestyle"). Within a given lifestyle group, individuals' decisions to purchase or use a product or service, for instance leisure air travel, vary significantly. These decisions appear to be those most open to the exercise of individual choice, i.e. subject predominantly to personal or individual-situational, rather than external macro, constraints. Personally constrained – but historically relatively un*re*strained – consumption behavior accounts for a significant part of continuously growing consumption rates, and often corresponding energy use, in household sectors like transportation and residential electricity use.

What might an individual's energy discretion continuum look like? This is much harder to produce, even as a theoretical exercise. The point at which discretionary becomes non-discretionary for any single person is murky, certainly variable and case-dependent, and seems to require the use of psychological or other specialized metrics to determine and characterize. A combination of influences – especially psychological, situational, and economic – shape what an individual regards as achievable and desirable consumption levels. Figure 1.4 offers a graphical representation of the lower portion of a hypothetical person's discretion continuum for personal transportation, with intervention types also indicated: it is easier to depict points on the y-axis in terms of concrete energy-relevant decisions or reduction measures. The order is highly idiosyncratic: another person's graph would look different. Note that direct behavioral measures to reduce consumption may appear at both the bottom and top end of the y-axis of personal discretion. For example, since this individual, for whatever combination of circumstances and preferences, places a high value on owning his own car, albeit a relatively fuel efficient one, foregoing ownership and renting when needed is a much less discretionary option for him: it therefore appears near the top of the y-axis at the less discretionary end.

Without recourse to well-defined disciplinary constructs, such an exercise quickly risks wading into murky philosophical waters, as in the distinction between a person's capability and his preferences to change behavior. The individual's and society's room to maneuver is explored further in later chapters: the transdisciplinary implications of both continua are important to sustainable consumption. This discussion finishes setting the stage for describing the approach taken in this book.

8. RESEARCH QUESTIONS AND CHALLENGES IN THIS BOOK

Confronting Northern consumption is vital for sustainability. This can only be accomplished with the aid and understanding of the public.[24] This book explores how advanced information tools can expand the public's

[24] The public's involvement is at least necessary if not necessarily sufficient.

involvement and participation in environmental decision making and deliberation.

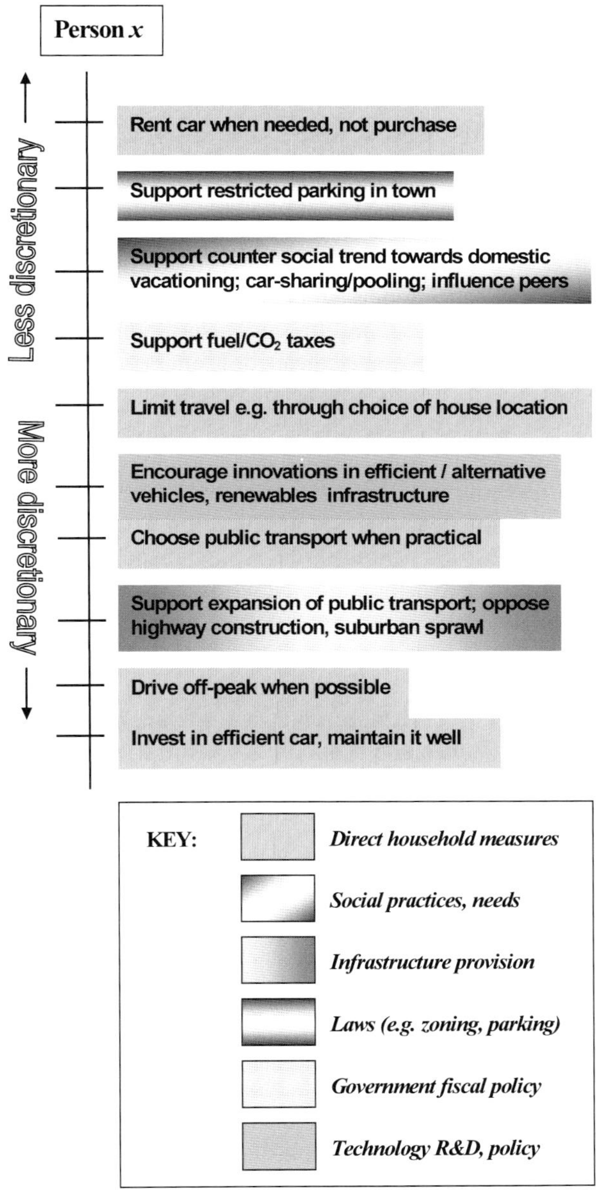

Figure 1-4. Excerpt from a hypothetical individual's discretion continuum for transportation, with intervention type (pattern key).

Environmental decision making is understood here not as the province of a few key environmental regulatory authorities or companies but in a broad sense as pertaining to a wide spectrum of technological, social, and economic conditions and developments that influence environmental sustainability. A number of recent or emerging theories of the environment and society, including ecological modernization theory, risk society, and reflexive modernization theory argue that a broader range of actors at different levels – including companies, consumers, social groups, and communities – should have larger and better distributed roles in spurring changes for environmental protection and reform [Mol 2000, Phung 2002].

To put this in an energy policy context, noting the small economic potential for energy efficiency improvements in a world of (recently) historically low energy prices, and the increasing tendency, especially in the United States, to leave decisions concerning energy technology and carbon emissions reductions to the discretion of businesses and energy companies under a regime of voluntary standards,[25] it is proposed that communities of end-users take a greater and necessary compensatory role in pressing companies and governments for environmental reforms. At the same time, despite the absence of sufficient economic signals and social mechanisms inducing personal restraint, householders, as responsible parties and key stakeholders, should also look to their own homes and practices for reduction potentials. How much of the one and how much of the other? Providing information and generating knowledge may help people find the appropriate balance and further lay involvement. A concrete result of this study is lessons for the design and use of improved information and communication approaches and tools for domestic energy conservation.

The assumption is that laypeople need better information on the array of factors and their interactions that influence energy consumption and by extension environmental sustainability. This aim requires different approaches to the provision of knowledge than conventional information approaches; and, as a testable hypothesis, it benefits from different communication tools. In contrast to the conventional provision of energy information, then, it is helpful to (let the end-user) confront the duality of, and examine the shifting border between, the discretionary and non-discretionary, e.g. between the behavioral and structural, in energy consumption.

[25] What some call a vacuum of leadership others describe as leaving matters up to the "free market," even though firms, particularly in the US energy sector, operate under conditions far removed from perfect, fully competitive markets.

What kind of information is best communicated to individuals to help them (i) conserve energy in their households and (ii) stimulate conducive structural changes to bring the household sector's energy consumption more in line with sustainability dictates? That is, assuming that an end-user-targeted, information/knowledge approach is constructive for this purpose – at least as an adjunct to directly targeting policy makers in government or technology supply – what sorts of knowledge sets (presumably comprising both scientific/technical and personal elements) are most important, and what might be useful and effective means of instilling, encapsulating, and communicating them?

An increasing number of studies of direct and indirect household energy consumption purport to enhance environmental awareness by providing information on the energy and emissions consequences of consumer behavior (see Chapter 3). Is this the right approach, or would it help to add information about the interaction of less discretionary driving forces? What kind of information and presentation actually best show the nature of such interactions and facilitate or motivate people to support some level of intervention to moderate their energy consumption, or their consumption in general? For example, are simple, non-differentiated indicators appropriate for most sectors of the public, or do some respond better to a more complex presentation of a complex reality? Can we treat energy like any other type of consumption and focus on the general forces underlying consumer society? To what extent?

Sustainable consumption at the household level is a viable target of study, and it may be partially amenable to intervention through information provision. But if so, it seems to require a more complex, multi-dimensional, and multi-disciplinary approach than is typically taken. I hypothesize[26] in this book that the individual, direct behavioral orientation in traditional energy information approaches and tools is an inadequate response to

1. the high levels of Northern (energy) consumption, as represented by Switzerland's 6000 watts per capita, an indicator of the burden of Northern consumption on the environment;
2. the complexity of ways in which the household sector in the North drives environmental problems and the transformations and innovations necessary to stem them; and
3. the untapped sources of lay intelligence, knowledge, and capability in apprehending the problems and contributing creatively to their solutions, which includes the potential for a richer scope and variety of involvement

[26] More specific research hypotheses are offered in Chapter 4.

of the public in confronting these problems and participating in processes of social and technological innovation that ameliorate them.

In general, this type of analysis and agenda has a close affinity to the sociological theory of the ecological modernization of consumption, especially its monitoring for enhanced consumer-citizen involvement at various levels in the production-consumption chain. This theory, as far as it bears on the research agenda, will be elaborated on in Chapter 3.

A broader accounting of the factors influencing end-users' energy-consuming devices, services, and practices may help facilitate their role in managing and shaping them. In the first place, it may induce personal or household conservation measures. In the second, a broader accounting might encourage end-users' contributions at positions higher up in the production-consumption chain. For example, it might enhance consumer-citizens' input into governmental and corporate technology policy choices or their support of environmentally favorable changes in social practices or norms. On a political level, such an information approach might also serve the aim of helping developed nations break free of the chicken-and-egg problem of mustering support from the electorate for institutional changes to restore signals for restraint (see Chapter 2, section 4.1). Expert modeling, in combination with exploratory discussions and interviews, is one way of providing such an accounting. This was pursued in the experimental portion of the study, as described in Chapter 4.

The next chapter prepares to develop this approach by expanding on many of the themes raised here concerning sustainable energy consumption. Chapters 2 and 3 explore theories and studies from several social science disciplines that cast light on consumption, information provision, and risk communication. All of this is crystallized and concretized in Chapter 4. That chapter describes the innovative redesign of a Swiss energy accounting program according to the theory developed in the preceding chapters as well as the program's use in structured, interactive interview sessions with householders. The software and sessions constitute a dual approach to energy-relevant technologies, behaviors, and social practices that allows end-users to make both a left-approaching "analysis of strategic conduct" (according to Figure 1-2) as well as a right-approaching "institutional analysis." The experiment provides a new look into people's understanding, and researchers' risk-communication, of the universe of factors that influence energy consumption along the discretion continuum. Chapter 4 also gives a full account of the experimental findings. The final chapter 5 integrates the lessons from the theories, literature, and field study to draw conclusions and point the way forward.

Chapter 2

TARGETS OF INTERVENTION FOR SUSTAINABLE CONSUMPTION

1. INTRODUCTION

Sustainable consumption, like sustainability and sustainable development, is notoriously difficult to define in a practical sense. The complexity, uncertainty, and normative dimensions and political divisiveness inherent in the global system sustainability is meant to characterize make it hard to agree on a set of suitable sustainability indicators. If the operationalization of sustainable development and the devising of suitable, universal sustainability indicators are two-way, cross-fertilizing activities [Verbruggen 1991], then given the state of the environment, it's no surprise that there is still much work to be done on sustainability indicators. In the case of sustainable consumption, the definitional difficulty has partly to do with the seeming impossibility to deal with a sufficiently large system of reference, even when focusing just on the economic and institutional component of the human-environment system. For example, encouraging the production and consumption of less resource-intensive but equally financially lucrative and satisfying goods or services is a start, but no firm or government will consider the scale of overall activity in the sector in its entrepreneurial calculations.[27] Environmental externalities are exactly about shifting costs to parties outside one's frame of reference, and short of an egalitarian world government and a universal "seventh generation"[28] ethical code, no one has an easy solution to the problem of the tendency of nations to develop at the expense of others' resources or companies to discount future generations' environmental stocks and quality. The difficulty also has to do with a resistance to look at the issue deeply enough, as in the taboo on differentiating between wants and needs for fear of violating consumer sovereignty. Indeed, it may be easier to say what is *not* sustainable about present consumption than to specify what would be sustainable.[29]

This chapter examines in more detail many of the factors driving environmentally significant consumption in the Northern capitalist-industrialist societies. Taken collectively, they characterize the current unsustainable system. They include market forces directly amenable to the most common interventions, including pricing mechanisms like energy taxes, technological innovations (especially in energy efficiency), combinations that encourage the purchase of 'green' products or electricity, as well as psychological means like exhortation (providing better information to consumers is saved for the next chapter). This is accomplished in the first two parts of section 2 in its review of patterns and levels of consumption. But overall, considering the poor record of the standard approaches (derived largely from economic and engineering disciplines) – especially taking a sufficiently large view of the problem and recognizing the special challenges posed by the current scale of environmentally significant consumption (the third part of section 2) – obliges us to move increasingly farther afield in search of driving forces, root causes, and practical means of addressing them, those forces and causes which in the last chapter we termed *less discretionary* from the vantage point of the individual. Accordingly, the discussion in section 3 ranges across the behavioral and social sciences, including psychology, sociology,

27 They are playing the economists' game of shuffling deck chairs on the Titanic, while no one is paying attention to Daly's "Plimsoll line," the line below which the boat must not sink if it is to remain afloat. Here the generalized sustainability indicator measures the distance between the actual water level and the Plimsoll line (and the unsustainability indicator measures how far below the line the boat has (temporarily!) bobbed).

28 The Great Law of the Iroquois Confederacy reportedly mandates that "In our every deliberation we must consider the impact of our decisions on the next seven generations."

29 "*Un*sustainability indicators," which measure unsustainable pressures and impacts, have been suggested for biogeophysical systems; it is often easier to describe and measure disturbances and imbalances than to characterize equilibria fully [Dahl 1995], [Munasinghe 1995b].

anthropology, economics, and ecological economics, in a review of theories, critiques, and recommendations from these disciplines that view (mis- or over-) consumption as problematic, mostly for environmental reasons but also for fueling various other societal problems.[30] As mentioned in Chapter 1, arguments and policies concerning the psychological, social, or economic problems stemming from consumption may also be brought for their potential usefulness in incidentally alleviating environmental consequences. The concluding section 4 wades briefly into political science to reflect on some political and policy lessons and implications. As an aid to the reader, short summarizing abstracts are offered after many of the lengthy subsections in this chapter.

2. EFFICIENCIES, PATTERNS, AND LEVELS OF CONSUMPTION

This section will discuss the three broad types of consumption and implied intervention points that have just been referred to: (i) patterns of consumption, suggesting qualitative changes in mixes of products and services; (ii) consumption of highly energy and material-intensive goods and services, pointing specifically to technological efficiency improvements (partly a subset of (i) and partly a supply-side intervention); and (iii) levels of consumption, suggesting, at least at the aggregate, a focus on quantitative reductions in material and energy throughput. Focusing on shifting patterns of consumption, along with necessary changes in production, is now common in research and policy circles; a much smaller minority insist that in addition, absolute levels of material consumption are too high in developed countries and must be reduced. A somewhat intermediate view is that in addition to changing patterns, exponential rates of increase in consumption of energy and materials should be lowered (since inflection points in the logistic (S-shaped) curves typically characterizing the evolution of the use of new technologies have not yet been reached in many cases), but not necessarily absolute levels.

2.1 Patterns of consumption

Agenda 21 implicates unsustainable patterns of consumption and production in global environmental decline [Symposium Sustainable

[30] An ethics treatment is also relevant but is not pursued for reasons of space and scope. See e.g. [Crocker 1998].

Consumption 1994]. To the extent that changing the forces driving unsustainable consumption patterns have been discussed in the environmental community, suggestions have focused mostly on technological innovation and correcting faulty price signals.[31] The general aim has been to change the type of consumption, not to reduce it. To this end, the United Nations Development Program, for example, has recommended promoting technological innovation, improving information and awareness, removing perverse price subsidies and adding ecological taxes, ensuring universal minimum (basic) consumption, and other related measures [Michaelis 1999].

Altering consumption patterns is a viable and necessary policy to reduce consumption's environmental impacts. In the United States in particular, industrial productivity increases since WWII have gone largely towards increasing income (rather than leisure), and a relatively large proportion of this expenditure has been devoted to the consumption of goods and services with a comparatively high material and energy intensity [Røpke 1999]. The reasons for this are complex and not fully understood, but it is clear that if the pattern of this expenditure could be directed towards lower-intensity goods and services, the environmental impact per dollar spent could be much reduced.

This is also the conclusion of a study on direct and indirect household energy consumption in the Netherlands [Vringer 1995]. The large differences found between the energy intensities of various consumption categories indicated that total household energy requirements could be reduced if the consumption mix were changed. The considerable variation among the total energy requirements of households within the same income category provided more support for this conclusion. An English study of the embodied energy of food in diets reached similar conclusions [Coley 1998]. Significant reductions in embodied energy are possible from a qualitative shift in diet (i.e. choosing alternative foods) without lowering caloric intake and, incidentally, without government-lead changes to agricultural, transportation, or retail practices. The implied assumption of both studies is that, provided the functionality and service remain the same, consumers can be encouraged to change what goes into their consumption baskets in an environmentally favorable direction. Given the necessary information and motivation, consumers could presumably initiate change on their own.

[31] This harks back to the classic environmentalist feud on the root of the problem in which Commoner emphasized technology and the economic system while Ehrlich and Holdren in their IPAT identity (Environmental Impact = Population * Affluence * Technology) considered affluence and population at least as important [Commoner 1971, Ehrlich 1972].

Expecting consumers to accept new diets as long as they provide an equivalent number of calories is unrealistic, and social and psychological theory (discussed in 3.2 below) has much to say besides about the forms and strength of non-utilitarian meaning that consumption has for many people. These studies are nevertheless helpful for analysts and policy makers (see also [Weber 2000]).

Table 2-1. The seven most environmentally harmful consumer activities in the US: Share of total impact[32]

Type of Consumption	**Global Warming**	**Air Pollution**		**Water Pollution**		**Habitat Alteration**	
	Greenhouse Gases	**Common**	**Toxic**	**Common**	**Toxic**	**Water Use**	**Land Use**
Cars and light trucks	27%	22%	46%	6%	13%	1%	13%
Meat and poultry	3	3	1	20	6	18	26
Fruit, vegetables, and grains	2	5	3	3	5	30	6
Home heating, hot water, A/C	16	11	5	3	1	1	0
Appliances and lighting	15	13	2	5	4	2	1
Home construction	3	4	2	6	6	1	23
Household water and sewage	0	0	1	11	0	5	0
Total	66%	58%	60%	54%	35%	58%	69%

32 [Brower 1999]

Similar effects can be achieved by substitutions in the broader product and service consumption mix. The American Union of Concerned Scientists used an economic input-output model to estimate the environmental impact of the goods and services American households typically consume [Brower 1999]. They found that the majority of environmental insult in the United States – in terms of greenhouse gases, air and water pollution, habitat disruption, and several other criteria – could be associated with only seven consumption categories: automobiles, meat, produce and grains, household appliances and lighting, home heating and cooling, home construction, and household water and sewage (Table 2-1). Automobiles and meat consumption dominated the list of environmental culprits, leading the researchers to conclude that even modest reductions in driving, driving cleaner or more efficient cars, or partly substituting grains and produce or even poultry for red meat could yield significant environmental benefits.

Patterns of consumption

Altering patterns of consumption in favor of less energy- and material-intensive products and services can ameliorate environmental effects and is a major thrust of mainstream environmental policy recommendations.

2.2 Efficiency and dematerialization

Promoting technological efficiency improvements is a primary concern in engineering and economics and is a mainstay of environmental policy. The theory and practice of increasing the energy and material efficiencies of manufacturing processes, equipment, and consumer goods and services are treated at great length in the literature; a summary from the latest World Energy Assessment [Goldemberg 2000] is offered here.

Improvements on current levels of efficiency are possible at many points along the chain of energy production and use, from extraction and treatment of primary energy resources, generation of primary energy, its conversion and distribution to produce final energy, final energy's conversion in turn to useful energy, and useful energy's transformation into energy services. This last step, involving end-use energy efficiency – reducing the energy needed (and lost) in providing services like lighting, cooking, and transportation – holds large but neglected potential for improvement through the development of new technologies. This potential's size depends on how the time horizon and boundary assumptions are specified. Variations in these parameters produce various potential terms like *theoretical*; *technical* (using best available technology, regardless of cost considerations); *market trend* (for a given year, assuming realistic prices and consumer preferences); *economic* (using the most energy-efficient technologies that are cost-effective, assuming well-functioning markets and the elimination of barriers); and others. Estimates of the economic potential for improving the efficiency of household energy service provision by 2020 – assuming year 2000 prices and constant utilization rates – are, in Europe, for new buildings: 20-30%; residential electric appliances: 35-45%; trains and railways: 20%; and aircraft: 25-30%. In North America some examples of economic improvement potentials by 2010 are electrical appliances: 10-33%; cars: 11-17%; railways: 16-25%; and aircraft: 6-11%. Estimates of the long-term potential by the end of this century exceed 80%, based on expected improved exergy[33] efficiency, use of new materials, increased recycling, substitution of natural raw materials for plastics, and so on [Goldemberg 2000].

In general it should be much easier, technically and economically, to implement options that increase the efficiency in the delivery of services than those that lower absolute material and energy consumption levels [Uiterkamp 1998]. Relative dematerialization has occurred in certain industrialized countries, driven by shifting consumer preferences for services, reductions in new infrastructure, increases in material efficiency, material substitution, recycling, and reuse [Goldemberg 2000]. Yet, in view of the technical potential and the need to ameliorate environmental effects, the general performance record is lacking.

Each of the new techno-economic paradigms to take hold in developed countries since the industrial revolution has led to a transformation of infrastructure, production, and consumption patterns and lifestyles. The current fifth paradigm, high technology and information, following the era of mass production and motorization from the late 1930s through the 1980s, may hold the technological potential for a relative dematerialization of the general economy. The necessary social and cultural changes in new paradigms are invariably slower to develop than the enabling technological innovations. Factor ten or greater efficiency improvements have sometimes historically occurred over short time periods, but generally only in response to strong technical or economic pressures and in suitable institutional settings [Michaelis 1999]. This may help explain the failure thus far to realize most of the dematerialization potential in industrialized countries. Other basic obstacles to improved energy efficiency include a poor perception of the potential by diverse and heterogeneous groups of manufacturers and con-

[33] Exergy: a measure of the quality of energy that, according to the Second Law of Thermodynamics, is degraded every time energy is used in any process.

sumers. The invisibility of efficiency improvements and investments is a liability for politicians who usually choose more image-enhancing measures. For their part, householders in Western countries – lacking knowledge, know-how, or technical skills – "under-invest" in energy-efficiency appliances or demand excessively short payback periods and high rates of return [Goldemberg 2000].

Real dematerialization of the economy has yet to materialize. In most cases increases in per capita consumption have outpaced increases in unit efficiency of materials and energy use [Goodwin 1997a]. The change from heavy industry to a service-oriented economy in the United States, underway in the 1970s, did not produce a reduction in the aggregate material and energy demand of the economy, in large part because services themselves were large resource consumers [Schnaiberg 1980]. Recently, American intensities of resource use, per capita or per GNP, are declining only for certain resources, and are increasing for paper, plastics, many chemicals, and, in several sectors, for energy [World Resources 1995]. The apparent decoupling of energy from GNP observed in the US after the 1970s energy crises may have in fact belied a continued close correlation [Tatum 1995]. Much of the recent spectacular rise in end-use consumption observable in affluent countries is material and energy-intensive, as seen from increases in such indicators as square meters of housing per person, number of automobiles and distances traveled, air travel, ownership of appliances, spread of air conditioning, and the like. Indicators for inputs and outputs like paper, electricity consumption and waste generation are generally moving in parallel with the indicators mentioned above and also support this conclusion [Røpke 1999].

2.3 Levels of consumption

A phrase adopted at a 1994 UN symposium on consumption has often been used as a working definition of sustainable consumption: "The use of services and related products which respond to basic needs and bring a better quality of life while minimizing the use of natural resources and toxic materials as well as the emissions of waste and pollutants over the life cycle of the service or product so as not to jeopardize the needs of future generations" [Measuring Changes 1998]. However succinct and convenient, each clause of this definition is problematic: (i) It is the satisfaction of continuously expanding wants-turned-needs, not basic needs, that is the object of a significant fraction of Northern consumption; (ii) the definition of quality of life is highly subjective and can be linked just as easily to consumerism as to environmental quality; and (iii) it is not clear that enhanced eco-efficiency and toxics reductions are sufficient to overcome the effects of constantly increasing consumption. The first two clauses are a subject of discussion in section 3 of this chapter, while (iii) is now discussed.

The preceding subsection asserted that large efficiency increases and dematerialization are certainly possible and may be highly desirable but that they have not occurred to a great extent. Yet, even if great strides were to be made in material and energy efficiency, would this be enough to contend with the environmental consequences of the consumption explosion?

[Measuring Changes 1998] questions whether eco-efficiency is sufficient to produce a reduction in total or per capita resource use. Spangenberg (1995), emphasizing the enormous size of the necessary reduction in resource utilization, insists that both an efficiency and a "sufficiency" revolution are required. A 1991 Dutch government study (cited in [Uiterkamp 1998]) concluded that additional measures will be necessary to maintain adequate environmental quality if production and consumption continue to increase as they have done. "If the technological options for adaptation become exhausted within the limited time available, then a fundamental revision of our expectations in regard to the nature and the extent of 'economic' growth will be unavoidable." Many others have noted the potential for rebound and scale to overwhelm eventually even the most efficient economies, given the growth in world population and the proliferation of resource-intensive consumer lifestyles [Whiston 1990], [Durning 1991], [Spreng 1994], [Harris 1997], [Michaelis 1999]. In affluent countries, efficiency gains in one area often stimulate demand for the product or service itself by rendering it less expensive, or they lead to the development of new products and areas of consumer demand [Michaelis 1999].

A simple use of Ehrlich's IPAT equation demonstrates that given projected population and consumption growth over the next 50 years, technological efficiency would need to improve by a factor of 16 to keep overall environmental impacts from increasing. If population and consumption growth were to be limited to the South, technology would only need to improve by some 79% over this time period [Ekins 1993], a target, as noted above, that comes close to the estimate for the long-term economic potential for energy efficiency improvements.

The scale of the macro-economy in relation to the natural world is of central concern in ecological economics and especially in the work of Daly.[34]

[34] Lintott notes the absence of consensus within ecological economics for the need for absolute reductions in throughput to reach a sustainable scale, although he brings good arguments in favor of such a consensus [Lintott 1998].

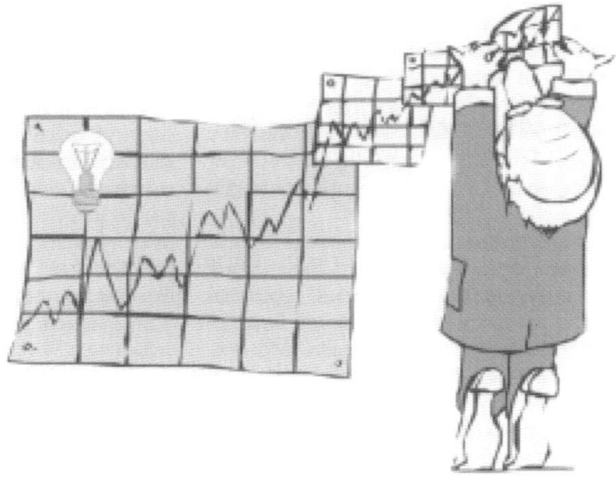

Figure 2-1. Illustration of continuous growth in electricity demand.[35]

Accordingly, aggregate national and global consumption must not be out of proportion to ecosystems' capacities to regenerate and absorb waste. When the global economy has reached an optimal size in relation to the biosphere, growth, defined physically (not economically) in terms of total material appropriation or throughput, should cease [Daly 1996a, b]. The actual maximum threshold level of throughput is a function of both ecological systems and technology and cannot be definitively determined because ofuncertainty and complexity. The precautionary principle recommends keeping throughput within some safe margin below the threshold value [Honkasalo 1998]. Development and maintenance of capital stocks, both natural and manufactured (not financial), at lowest cost should then be the goals of economic activity, producing an economy in steady state with respect to this capital [Daly 1996a, b] (see also [Jaeger 1994]).

The evidence for the swamping of efficiency increases is clear to many: "There are few, if any, examples where attempts to 'save' energy or resources actually led to an overall reduction in the *total use* of that energy or resource" [Princen 1998] (emphasis added). Whiston (1990) asserts that the increasing scale of Northern consumption and the South's emulation mean that conventional pollution control and energy efficiency only chase a moving target. Similarly, Daly claims that the severe environmental degradation in many places and across many media provides good evidence

[35] Source for illustration: [Schwarz 1996].

that the global economy has already reached, and probably surpassed, its optimal scale [Daly 1996a, b].

The directors of the Dutch HOMES study on household metabolism[36] admit that increases in the efficiency of service delivery alone will probably prove insufficient to reduce sectoral household metabolism in the Netherlands, forcing consideration of means to reduce unit household consumption levels. They note that on a sectoral level, technological progress has been offset by large increases in population, the number of households (driven largely by socio-demographic trends towards smaller households), incomes, and investments in distribution grids for access to energy and water. Indirect household energy consumption has been growing irregularly since the 1970s. In the 1970s this increase seems to have come about from growth in the volume of general end-use consumption. Growth leveled off after the 1979 oil crisis, as did indirect energy requirements. After 1984 the volume of consumption rose again along with indirect energy demand. Notably, change in the structure of consumption patterns (e.g. shifts in purchase trends from one production sector to another) did not significantly change total indirect energy demand [Noorman 1998b, van der Wal 1998, Wilting 1998]. Thus, in the environmentally progressive Netherlands, both efficiency increases and changes in patterns of consumption have been wholly inadequate to contain the growth of household energy use and the implied environmental consequences.

A study of the recently developed Spanish Valencia [Garcia 1998] [New Consumers 1999] brings home the inadequacy of focusing on patterns and efficiencies alone:

> Valencians became "new {Northern} consumers" at the end of the seventies, after a period of intense industrialization. Since then, the model of consumption {homogeneous mass consumption and consumerism} has remained basically unchanged, but it has increased its social and environmental costs. In its capacity to meet needs, it does not differ much from the 1980s system. However, it requires a growing (physical and economic) effort to keep "fit". ... It needs more power, produces more waste, and exerts a greater pressure upon the ecosystems.

The implied need for increased sufficiency, not just better efficiency, is examined in detail in the subsection on economics in the upcoming section 3.

In terms of energy supply and demand, the continuous increase in the global scale of energy use also means that technological solutions that focus

[36] Household metabolism: Flows of energy and resources through the residential sector.

only on the development of alternative energy sources will likely prove inadequate. Despite rapidly increasing investments in photovoltaics, wind power, biofuels, and hydrogen-powered fuel cells, technical and economic hurdles (especially for hydrogen) combined with the enormous and ever-growing scale of world energy use (especially for transportation) imply that these climate-favorable energy sources could at best meet only a small fraction of world energy demand for quite some time to come, perhaps decades. This requires placing at least equal emphasis on restraining energy consumption through the mid-term.

Efficiencies and levels of consumption

Large improvements in energy efficiency and a relative dematerialization are certainly possible and may be highly desirable, but the current state of implementation falls greatly short of the technical and economic potential. Evidence suggests that even such improvements would be inadequate in the face of the increasing scale of the Northern (and global) consumption-based economies.

3. AN EXPANDED MULTI-DISCIPLINARY ANALYSIS OF CONSUMPTION'S DRIVING FORCES AND POSSIBILITIES FOR INTERVENTION

This section considers driving forces, issues, critiques, and policies related to consumption from a wide variety of disciplinary and multidisciplinary perspectives in the social sciences, including behavioral, psychological, social, and economic. The material is grouped by discipline and critique, an approach partly suggested by the extant literature.

Expanding consumption has been the subject of research in all social science fields and in many humanities, yet on the whole the diverse theories that have been produced have not proved robust enough to explain or predict the phenomenon very well [Wilk 1999]. The lack of an adequate interdisciplinary conceptual framework has also hindered the development of effective policies for sustainable consumption [Michaelis 1999]. Alternative consumption theories have also not been subjected to much empirical testing [Schor 1995].

Household consumption, and especially associated rates of material and energy efficiency, are functions of biophysical, technical, economic, spatial, and behavioral factors as well as specific social institutions and administrative policy measures. The constraints of each of the domains in which these factors operate determine the potential for changes in household consumption. As the HOMES researchers put it: "Designing effective and socially acceptable policy instruments to reduce the rates of household metabolism (and their negative environmental impacts) requires a thorough understanding of the determinants of household consumption and their mutual relationships, as well as detailed information on possible differences between the 'lifestyles' and resulting consumption patterns of different segments of the population" [Noorman 1998b].

Households' behavioral and social functions are dynamically linked to physical and social structures and these links show varying time delays, time horizons, and spatial connections [Uiterkamp 1998]. For example, the Needs-Opportunities-Abilities (NOA) model of individual consumptive behavior is embedded in a societal context of technology, economy, demography, institutions, and culture (the TEDIC complex) [Gatersleben 1998]. At higher levels, still broader matrices are necessary: "The environment and ecology problem at both the national and global level {is} the outcome of complex socio-historical, valuative, and cultural developments. Any lasting global solution has to recognise that underlying matrix" [Whiston 1990].

Disciplinary approaches and analytical possibilities

Numerous, diverse theories of consumption have yet to prove robust enough adequately to explain or predict the phenomenon, in part because of the lack of an adequate interdisciplinary conceptual framework. Recently, interdisciplinary approaches to household consumption have attempted to incorporate biophysical, technical, economic, spatial, and behavioral aspects, as well as specific social institutions and administrative policy measures affecting household consumption flows.

3.1 Consumption critiques

Many insights into consumption and innovative policies for ameliorating its environmental effects come from critiques of various aspects of the consumer society: "A critique of non-ecological aspects of social organization is fundamental to every kind of environmental protection" [Cogoy 1995]. This section gives an overview and sample of the approaches and types of critiques across disciplines in the literature, which are then examined in greater depth in later sections.

As noted above, there is a "divided discourse" on consumption from isolated academic departments, making results limited, difficult to communicate outside the discipline from which they arise, and embedding assumptions of the discipline in the research design. Both academic proponents and critics of the consumption basis of capitalism take a moral position on what is good or bad for society "ontologically prior" to their research on consumption [Wilk 1999]. Thus, many theories contain implicit embedded critiques. Wilk insists that many of the diverse theories of consumption can be "correct" or applicable under the right circumstances. He calls for the "develop[ment of] meta-theoretical guidelines specifying which models are useful in which empirical situations," resulting in a "heterodox multigenic theory, which accepts that there are multiple determinants of consumption..." [Wilk 1999].

Schor suggests four theoretical bases for critiques of consumer society [Schor 1994]:

1. Schor's own approach (described below in 3.4.4, "Break the work-and-spend cycle"), which stresses a market failure in which workers may not choose the length of their work time and opt for greater leisure time and must instead work long hours for higher salaries. These accustom people to high and often continuously increasing consumption levels.
2. The failure to price natural capital and environmental externalities adequately, and other criticisms from ecological economics.
3. Resulting damage to community and social values such as integrity, honor, responsibility, trust, caring, and sharing.
4. Effect of social interaction on consumption; e.g. positional consumption means that aggregate growth in consumption or incomes can never increase overall economic welfare or happiness (described below in 3.4.2, "Decoupling consumption and welfare").

A grosser classification divides critiques into two camps, those that emphasize one or more of the various underestimated costs of consumption or over-consumption (critiques two and three in Schor's taxonomy) and, secondly, those that question the extent of presumed benefits of that consumption. Since costs may be seen as dis-benefits, these two types converge at some level of analysis. Practically, however, the two approaches emphasize very different things.

The traditional critiques of consumption and consumer society "stress the costs of consuming – in terms of environment, time, community, and quality of social interactions" [Schor 1994]. Moral critiques add ethical or spiritual costs. More recently critics, especially practitioners of ecological economics, have suggested that the environmentalist case against consumption rests on a much firmer base if it questions the benefits of consumerism (de-couples economic welfare from consumption), rather than just emphasizing the costs. In addition, the environmental costs are subject to uncertainty and therefore popular rejection and political manipulation [Lintott 1998]. For instance, modified (e.g. green) consumption still holds consumerism as the goal, accepting its implicit assumption that it goes hand in hand with welfare. This implies that if sustainability requires a reduction in consumption to benefit society in the long-term, it must reduce welfare in the short-term, a politically difficult goal. Reductions in consumption are much more plausible if the consumerism-welfare connection is questioned [Lintott 1998].

Some researchers have sought to combine consumption policies with employment and environmental issues with which consumption is intertwined: The "three-legged stool" of arguments in favor of less consumption, fewer working hours and less environmental damage combines a critique of consumption, an environmentalist critique, and a critique of current employment-related issues: "Combined with income and work sharing, this could be a strategy for simultaneously dealing with problems of environmental cost, welfare, and unemployment and poverty, in a way that current policies have failed to do" [Lintott 1998]. Other types of critiques also implicitly or explicitly involve several aspects of society.

Consumption Critiques

Many insights into consumption and innovative policies for ameliorating its environmental effects come from critiques of consumer society; yet critiques and assumptions embedded in consumption theories can also stymie interdisciplinary communication and cooperation. Critiques can be grossly classified as those that emphasize one or more of the various underestimated costs of consumption or over-consumption and, perhaps just as important, those that question the extent of the presumed benefits of that consumption.

3.2 Psychological and behavioral treatments

3.2.1 Studies of behavior

Although consumption is in some way a part of most human behavior, and behavioral and psychological mechanisms underlie household consumption patterns, the behavioral underpinnings of consumption are largely unknown and unexplored [Princen 1998]. Moreover, since it is defined largely with respect to a system's capacity (e.g. the biosphere), the problem of *over*consumption is a specifically meso- or macro-phenomenon. Within systems of appreciable size, overconsumption does not have a definite meaning at the individual level, since the effects of individuals' consumption behavior are clear only in the aggregate [Princen 1998]. Its inverse concept of sufficiency, examined below in the economics section 3.4.3, "Restraint and sufficiency," works at the level of the individual or small group.

The "Needs-Opportunities-Abilities" (NOA) construct mentioned above is an example of a recent conceptual model of consumer behavior. Opportunities represent factors and conditions stimulating consumption behavior while abilities such as financial means are constraining factors. Needs and opportunities feed into the motivation for consumption, while opportunities and abilities establish the "behavioral control" necessary for the purchasing action [Gatersleben 1998].

Empirical studies of Dutch household consumption behavior trends have also produced interesting findings on the now often studied relationship between environmental awareness and consumption behavior. Most Dutch consumers, it has been concluded, are well aware of the environmental impacts of their behavior – at least those stemming from the use of products (e.g. direct energy expenditure), if not their production and acquisition (grey energy) – but this understanding does not motivate them to change their consumption behavior. Moreover, most survey respondents agree that it is necessary and even possible to change many household consumption behaviors, except for reducing driving (necessary but only slightly possible) and living in a smaller house and taking vacations closer to home (neither necessary nor negotiable). Those who are aware that they are inflicting relatively greater environmental damage are not more willing to change their behavior than others. *In short, the Dutch show neither gross ignorance nor denial of their role in environmental damage but a combination of inability (e.g. the car) and unwillingness (e.g. far-flung vacations) to change their behavior patterns* (to "cramp their lifestyles") [Gatersleben 1998].

In the early 1990s, sociologist Lutzenhiser reviewed the literature on behavioral and social aspects of household energy use [Lutzenhiser 1993]: The reader is referred to his study for detailed references. He called for an interdisciplinary, open-minded effort to develop "an over-arching model that can simultaneously capture group dynamics, body use, cognitive processes, and human-machine interactions." Some of the more interesting studies he reviewed investigated patterning of behavior and energy use; self-awareness and accounting for energy consumption; bi-directional influences of consumption behavior and attitudes; and differential adaptation to price increases across different family types.

Specifically, behavior and end-use consumption are highly patterned and therefore stable over time periods within a single household. Bernard (1988) proposed classification of household energy use as (1) structural consumption that occurs when the building is unoccupied; (2) habitual consumption that occurs from routine conscious and unconscious management; and (3) daily variable consumption from unusual events like holidays, vacations, sick children, and visitors [Bernard 1988]. Unconscious habits play an important role in energy use, although little research on them has been conducted. Habitual, unconscious behavior, uncertain consumer knowledge, and endemic reporting errors weaken conventional energy analysis.

The "limits of human cognition" may prevent an accurate self-assessment of consumption practices. For example, average rather than marginal costs are usually conceived in personal energy accounting.

Linz and Heberlein (cited in [Lutzenhiser 1993]) found that once certain Midwest American household residents had gained experience with peak electricity rates that were quite high compared to non-peak rates and with peak periods that were as long as twelve hours, they developed a sense of social obligation to shift consumption away from peaks. Thus (economically driven) behavioral changes can produce new energy attitudes, rather than conservation attitudes necessarily preceding behavior. Of this Lutzenhiser says that energy rates and billing arrangements are embedded in, and come to be accommodated by, pre-existing social institutions.

Studies of behavior

The behavioral underpinnings of consumption remain largely unexplored in research. A recent study of the connection between environmental awareness and consumption found that the Dutch show neither gross ignorance nor denial of their role in environmental damage but a combination of inability and unwillingness to change their behavior patterns. Other studies investigate patterning of behavior and energy use; self-awareness and accounting for energy consumption; and bi-directional influences of consumption behavior and attitudes.

Finally, different types of families were found to respond differently to rapid energy price increases in the US. Lower income families made lifestyle cutbacks across nearly all end-uses; higher income households maintained consumption and/or used tax credits and incentive programs to invest in energy-efficiency improvements in buildings and equipment.

3.2.2 Needs and wants, identity and meaning

Consumption in consumer societies of the North is often contrasted with consumption in less developed economies, a large portion of the populace in the latter said to be preoccupied with satisfying "basic needs," and those in the former with fulfilling ever greater "wants." However, beyond the level of absolute subsistence, the psychology of needs within societies and the connection between needs and wants is much more complex than such pat contrasts suggest.

The relationship between economic consumption and the satisfaction of needs is often exceedingly complex and non-linear [Jackson 1999]. Many different theories of needs have been proposed, including hierarchical needs theory and Rokeach's (1979) theory of eighteen instrumental and eighteen terminal values. The latter has been related to means-end theory in describing and defining consumer behavior and consumer lifestyles [Gatersleben 1998].

Many critiques of consumption associate its huge increase, particularly in certain categories like durable goods, with attempts, often largely unsuccessful, to fulfill through consumption various underlying psychological needs that are no longer being fulfilled through other social forms and mechanisms. Advertisers play to these underlying needs by associating products like cars with social status, sex, personal power, recreation and leisure, and freedom and creativity. Mobility itself is not a need; people travel in order to fulfill other needs, like subsistence, protection, participation, affection, and freedom [Jackson 1999] (although travel can itself become a pleasure or "want," as some of the interview subjects testified (Chapter 4). Clearly, economic theory to the contrary, consuming market goods and services is not necessarily the same as satisfying human needs and wants; it is only one culturally specific means to that end, and not necessarily always the most effective.

"The concept of needs, their relationship to desires, and the ethics of satisfying or denying them, is a subject of ongoing and perhaps irresolvable debate in the social sciences" [Michaelis 1999]. The relationship of luxuries to necessities, and the transformation of the one to the other, is an important area for consumption and sufficiency research. Goods newly on the market are often considered luxuries, while older goods have been deemed necessities [Schor 1995]. Yet the longer and the more any such goods people possess, the more important they deem them for maintaining their quality of life [Gatersleben 1998]. [Wilk 1999] has much to say about the relationship and process of transformation between needs and desires in a social context (presented below in section 3.3.2, "Anthropological/sociological treatments").

Neoclassical economics asserts axiomatically that wants are insatiable as a whole, i.e. that the accumulation of wants has no in-built limits. Yet sociology, anthropology, psychology, and Eastern philosophies take it as fundamental that wants are pliable and that therefore "scarcity" can be a function of the level of wants [Goodwin 1997b]. Leiss challenged the notion of insatiability on psychological grounds: People often do not know if and how goods satisfy their wants; thus they cannot continuously generate new wants [Leiss 1978].

In the course of evolution of the prevailing economic development paradigm, consumption has come to rival or in some cases replace work as a source of individuals' identities. The blossoming of mass consumption and advertising after WWII was important to solidifying this transformation. Consumption thus acquired a new and important function in the bestowal of self-identity. With increasing individualism and the loss of community bonds and well-defined social roles comes the necessity for the individual's sense of identity to be forged actively. For many, consumption plays a role in the process of building and maintaining identity. Similarly, to some extent people use consumption as a substitute for the reduced meaning they find in work and other activities, affiliations, and institutions [Kiron 1997]. Others claim that this process is exaggerated and that social contexts and other factors come into play as well. For instance, [Michaelis 1999] points out the establishment of ties to specific communities of consumers as important with the erosion of more traditional communal associations. Overall, the identity, sense of community, and other forms of meaning that consumption provides in a culture or era otherwise seemingly lacking in these things deeply entrenches consumption in the psychological fabric of Western life. It is thus all the harder to dislodge.

The practice of positional consumption, consumption done in competition with others in one's class or group and in emulation of those above for the satisfaction of "relative needs," is said to account for a significant portion of consumption in the US and other Western countries. It is explored in the social sections below.

This chapter does not explore the fields of environmental awareness and public attitudes (leaving that for Chapter 3), but in connection with the discussion of material wants, the debate on materialist/post-materialist evolution is worth mentioning. Through generational cohort analysis, Inglehart concluded that generations who grew up under difficult economic conditions and deprivation become more materialist in their values, while those generations who grew up in post-war prosperity are more non- or "post-materialist." Dominance of attitudes among generations makes the pace of change of materialistic values quite slow. Cross-sectionally, post-materialist values are more highly associated with higher occupational status and higher income. Yet economic conditions have some affect on the strength of these values. Long-standing economic downturns fuel the growth of materialist values, and upturns their relative inhibition[37] (as cited in [Schor 1995]).

Inglehart's post-materialist theory shares the common conclusion of the literature on "environmental modernization." There, the most modernized groups, those with higher economic and educational levels, are presumed more likely to develop post-materialist, environmentalist values. Yet, there is empirical evidence that whatever their values, the relative environmental impacts of the consumption lifestyles of the more modernized – e.g. the demands on the infrastructure of their diet, housing, mobility, and recreation – are often greater than those of lower classes or the less developed, pointing to Garcia's "anti-environmental modernization" described above [Garcia 1998], [New Consumers Conference 1999].

Needs and wants, identity and meaning

The psychology of needs within societies, the connection between needs and wants, and the relationship between economic consumption and the satisfaction of needs are exceedingly complex. Many critiques of consumption associate its huge increase to attempts to fulfill through consumption various psychological needs that are no longer being fulfilled through other social forms and mechanisms; consumption may provide identity, a sense of community, and other forms of meaning. If wants are pliable or satiable, scarcity is largely a function of the level of wants.

3.3 Social treatments

3.3.1 Culture, society, and lifestyle

Society and culture are important influences on consumption of all kinds. As demonstrated in historical and psychological work, consumption and consumption growth have become deeply embedded in cultural norms and values, such that many people's sense of well-being, and others' perception of them, depend substantially on their earnings and possessions [Gatersleben 1998]. The encouragement of uninhibited consumption is a salient feature of the prevailing Western culture (the "culture of capitalism") [Robbins 1999].

Cross-cultural studies of materialism reveal systemic cultural differences in the meanings ascribed to consumption. Interestingly, although all cultures officially denigrate or condemn excessive materialism, paradoxically

37 This seems self-contradictory, since upturns are partly related to more consumer spending and confidence, suggesting perhaps a type of staggered negative feedback loop between macro-economic conditions and the strength of materialist values.

members of all societies have high consumption aspirations. Individuals view their own consumption as good and justify and legitimize it in various ways [Røpke 1999].

Culture is responsible for resource constraints in an important way, and not just as a result of actual anthropogenic depletion or degradation. "The threat of scarcity is a socially manufactured, permanently entrenched characteristic of any society that connects the satisfaction of needs to consumption of goods, and this threat will not be diminished by increases in the supply of goods. {Yet} the threat of {actual} scarcity has returned as a significant economic issue as mismanagement of industrial waste products begins to pose global environmental threats" [Leiss 1978].

Sociologists and anthropologists insist that technological systems[38] and the patterns of life they serve are highly socially constructed or shaped [Lutzenhiser 1993, Tatum 1995]. Yet once those systems and patterns are in place they tend to become entrenched (sometimes until some point after the next major technological revolution), and individual choices are then to a lesser or greater extent bounded by the socio-technical framework. Individual new commodities or consumption fads become integrated into internally propelled systems of commodities, physical infrastructure, social practices, and institutions and within technological trajectories [Røpke 1999]. If they reach a certain critical mass and fit the existing system well, they induce further modifications of the system to serve them. Alternative consumer choice becomes scarcer and increasingly difficult as the technology or consumption pattern becomes "locked in" (see also [Unruh 2000]). The automobile is a classic example. Change requires the penetration of an alternative, often in the service of a different social need [Michaelis 1999].

Durning, an economist who finds fault with the consumer economy for environmental reasons, holds that its remedy is to be found in such cultural change [Goodwin 1997b]. There is a large cultural component in the definition of success, and it is probably larger the more the national consumption exceeds subsistence levels. Similarly, well-being, above a basic level, is largely open to cultural definition. This definition can be changed, especially if it becomes dangerous [Goodwin 1997b]. If the proper social structures can be created, betterment can be (re)defined in non-material, perhaps qualitative ways [Lintott 1998].

[38] Systems researchers like Hughes use the term *technical systems* to refer to systems comprised exclusively of technological (inanimate) objects, *technological systems* for combinations of technical systems with human actors, and *socio-technical systems* (or techno-economic or other variations) for large, complicated technological systems such as cities or even cultures [Hughes 1997].

In local, social groups of ecologically-minded individuals (like Gershon's Eco-teams[39]), people's interaction and engagement with co-members has been found to lead to reflexivity on an individual and collective basis: Membership in such a community leads to self-reflection on preferences for consumer goods [Georg 1999]. The feasibility or potential of such local social "islands" for inducing society-wide change is discussed in section 4.1.

Most prominently in the United States, prevailing cultural norms are heavily weighted in favor of quantitative consumption, particularly of the competitive, positional sort. The pursuit of material goods in status-based competition with others for the latest item on the market or the most exotic vacation guarantees non-satiability: More purchases are always necessary to keep up with the "Joneses" on the latest trend, which leads to endless "ratcheting up." Surveys show that the amount of money thought needed to sustain an average-sized family in reasonable comfort in the US (adjusted for inflation) has consistently risen over the years, and is always close to income ("needs" follow income), but is always considered somewhat over the amount the median family actually has, all suggesting positional consumption is at play [Schor 1995]. This results in chronic dissatisfaction and frustration, as well as ever-increasing environmental stress, at least assuming constant environment impact per dollar spent.

In *The Overspent American*, Schor argues that the great majority of (US) Americans now aspire in their spending styles to those of the wealthiest upper quintile of society. While it used to be that people strove for the standard of living of the next one or two income levels above them, now the reference group is those with incomes three to five times their own. Large levels of consumer debt is one result [Schor 1998]. Television and films fuel Americans' high consumer aspirations, not only through advertisements but also the lifestyles of the rich portrayed on the screen [Uchitelle 1998]. Robert Frank's *The Winner-Take-All Society* and *Luxury Fever* share many of Schor's other findings, but suggests that consumption emulation is still more limited to the consumer levels just above the individual's level[40] (cited in [Uchitelle 1998]).

Human behavior in a social context is apparently just as important in energy consumption as it is for end-use consumption in general, although it has been neglected in traditional energy analysis [Schipper 1997], [Lutzenhiser 1993, 1997]. Contrary to the assumptions of the dominant

[39] See section 5.1.3 in Chapter 5 for an elaboration on Eco-teams.

[40] Brooks' recent investigations would suggest that Schor's sort of high aspirations are much more characteristic of the "Blue" half of America than of the "Red" half. See [Brooks 2001].

physical-technical-economic model, residential energy use and associated practices of residents vary tremendously with cultures, social networks, communities, and families [Lutzenhiser 1993, 1997]. A US Department of Energy study of changes in household energy use in the US from 1979 to 1987 suggests that social and behavioral factors played an important part in restraining sectoral energy consumption over that period. Differences in gender, division of labor, and social work roles have been found to play a part in energy conservation attitudes and energy-relevant activities [Lutzenhiser 1993].

Lutzenhiser and Hacett's study of reactions to changing the incentive structure in residential apartments from master metering to unit metering found variations in consumption as large as 300% among very similar apartments, due to differences in family size, length of residence, income, ethnic culture, and the like. On a building level, consumption varied considerably between two neighboring complexes otherwise nearly structurally and socially identical, reflecting "locally evolving standards of behavior" [Lutzenhiser 1993]. Surveys showed that households that consumed significantly less energy in response to changed signals did not do so because of stronger conservation ethics or environmental sensitivity. "Rather, they had different ethnic and cultural manners of behavior and household organization. Thus cultural practice and collective restraint can produce highly variegated and lower-than-expected consumption levels among households that by economic reasoning should be likely to exploit common property resources"[41] [Lutzenhiser 1993]. Investigating how to reproduce the conditions for this group phenomenon over a wider area is a crucial part of restraint and sufficiency research (examined below).

For decades energy analysts have favored the term *lifestyle* to describe different patterns of behavior and consumption characteristics of various social groups. In one of its best usages, lifestyle is similar to the anthropological concept of culture: the "totality of practices, meanings, beliefs, and artifacts of a social group," [Lutzenhiser 1993] although the term and concept have been used in many different ways. Lifestyle analysis has the advantage of capturing certain aspects of social behavior not well defined by traditional economic or demographic means. For example, recent lifestyle studies in energy research have examined groups defined on the basis of common clusters of household composition, hardware, activities,

[41] The notion of (studying) restraining consumption through (explicating) social or cultural means, rather than just by showing environmental costs and consequences in order to cultivate environmental consciousness, is posited in the Social-Revealing approach, discussed in Chapter 3.

schedules, and beliefs. Lifestyle groups so defined cut across more than one demographic grouping (age, income, ethnicity) [Lutzenhiser 1993].

While related in certain ways to psychological and other micro-behavioral studies, most lifestyle-type research has been conducted by social scientists and marketing researchers. Social scientists account for consumption differences among demographic categories by appealing to class and sub-cultural differences and constraints. Marketing researchers identify clusters of consumer characteristics and attitudes associated with differences in purchasing behavior, i.e. lifestyle-based market segments. A major weakness of lifestyle segmentation schemes is that they are mostly descriptive. They do not help to establish how lifestyles are created or how the relationships between lifestyles and the social structure are shaped. Lutzenhiser suggests combining social science and marketing approaches in lifestyle analysis to explore lifestyle origins, freedom of choice, dominance of certain lifestyles, boundaries, and possibilities for change with the aging of cohorts and general socio-economic change [Lutzenhiser 1993]. More recently he has highlighted the growing importance for energy analysis of the system of status-graded lifestyles dependent on social class, which seems to be predominantly a function of wealth: The poorest consume energy only modestly, while the wealthy tend to use large amounts of energy [Lutzenhiser 1997].

Income is strongly positively associated with the consumption of resources, as measured by indirect energy, for example. Yet there seem to be non-linear, semi-qualitative thresholds between the social classes in terms of the environmental effects of their respective lifestyles: The environmental impact of the lifestyle of the higher groups seems to be disproportionately larger than would be predicted simply by their greater spending ability [Garcia 1998].

Vringer and Blok [Vringer 1995a] studied differences in total energy requirements across income groups in the Netherlands. In general, there are appreciable differences in energy intensities within and among household consumption categories. Moreover, for each group, the energy requirements of the majority were found to be within plus or minus 25% of the average value. They took this figure as an estimate of the potential for short-term reductions in household energy expenditure. This deviation from the average is similar at any income level (ceteris paribus), suggesting that on top of the income driver is another uniformly acting factor. This may be a "lifestyle" factor [Wilting 1998].

In general, many researchers consider the study of emerging and diverging lifestyles and their relative impact on resource use to be the next important step in household research [van Diepen 1998]. As described in Chapter 1, the general gravitation towards more individualistic lifestyles since the 1950s has driven such environmentally significant trends as smaller households, larger and better-equipped houses, and increased car ownership.

Culture, Society, and Lifestyle

Technological systems and the patterns of life they serve are highly socially constructed; yet individual choices become bounded by entrenched socio-technical frameworks. If the proper social structures can be created, success, well-being, and betterment can be defined in less material ways than at present. Positional consumption tends to foster a sense of non-satiability. Human behavior in a social context is also important in energy consumption. The environmental impact of the lifestyle of higher income groups seems to be disproportionately larger than would be predicted simply by their greater spending ability; however, variable lifestyles within single income level groups also seem to account for measurable differences in consumption intensities.

3.3.2 Anthropological/sociological treatments of needs, wants, and restraint

The discussion now returns to a consideration of necessities and luxuries in the context of sociological and anthropological treatments of needs and wants. A social context, rather than a purely individual psychological perspective, permits a deeper understanding of the societal transformation of wants to needs and their embodiment in norms, as well as a view to the social and technological constraints on individual choice and on attempts to hold "needs" up to question.

The expression of individual and social identity in terms of cultural consumption rather than experiences related to productive activity has been mentioned previously. The growth and expansion of leisure and "luxury" consumption, and corresponding tastes and expectations, have become particularly pronounced in recent years. In developed countries the luxury goods market has gained in size at the expense of cheaper goods as wealth disparities have increased. The market for fuel-intensive sport utility vehicles in the United States, for example, has quickly grown to represent half of all new vehicle sales.

Wilk's anthropological theory of needs and wants provides a plausible explanation for the growth in "luxury" consumption and suggests how it might be modified [Wilk 1999]. According to Wilk, consumption is a complex balance between diverse forces that are multigenic and dynamically, multiply, and complexly linked. Consumption also involves constraints, restraints, and limits, most of which, except for income, are not recognized in the literature. Including restraints or inhibitions recasts consumption as a balance of needs and wants between these inhibiting factors and other promoting factors.

Wilk conceives of *needs* as accepted social standards of living, whereas *wants* are consumption desires that are generally considered beyond those standards. Needs need not, and in developed countries for the most part do not, conform to the definition of basic human needs. In modern societies, wants are commonly, and now increasingly rapidly, transformed into (perceived and accepted) needs, and consumption is the process of filling them.

Social standards of consumption, like other social rules, are taken for granted and not questioned. They can only change if they are brought from the realm of the "unconscious habitus into the discursive sphere of heterodoxy where they are subject to manipulation, evasion, and multiple interpretation" [Wilk 1999]. That is, needs may follow a cycle in which they are consciously questioned or challenged (moving from the habitus to heterodoxy), and then re-framed as "wants." If the challenge is successful, the needs are identified as wants and placed out of the realm of universally socially accepted standards. If not, they become reestablished in the habitus as legitimate needs.

Two processes control the rate of expansion of needs and the production of new needs. Naturalization consists of forms of social control that maintain the status quo by keeping needs inscribed in the habitus and not letting them be questioned, asserting that the existing order is natural and that other practices are unthinkable. Cultivation is the "opposite process of opening existing needs to question, discussion, and debate" [Wilk 1999].

Changing societal behavior thus involves changing the habitus, inducing reconsideration of one or more of the practices that are daily taken for granted and never questioned by bringing them into the realm of discourse where "needs" can be reconverted to "wants." This is what policy tools must target that seek long-term solutions to the environmental problems from consumption. Tools and strategies must follow the cycles of cultivation and naturalization. Unfortunately, Wilk notes, "at this point we have little systematic knowledge of the social, psychological, and communicative practices that naturalize new practices and understandings" [Wilk 1999].

This kind of targeting holds particular promise for discretionary[42] energy consumption, for example not that which is largely unavoidable due to existing technologies and infrastructure (e.g. the need in most cities in the US to use an automobile for shopping and commuting), but consumption that at least until recently was regarded as a luxury, like driving in sport

[42] See Chapter 1.

utility vehicles or flying often to far-off vacation sites. An example of bringing an established consumption behavior out of the habitus back into heterodoxy is the anecdotal development in parts of Scandinavia in which it has come to be considered socially stigmatizing to drive alone in an auto to a neighbor's house for a social call.

Along these lines,[43] Georg notes that, in general, "environmentally informed modes of behaviour" have not spread into the mainstream (habitus) of the public, financial institutions, and governments. "Norms for environmentally sound behaviour appear to be at a stage of pre-institutionalisation. Whether these will be institutionalised as a norm depends on whether a broader consensus about the importance of these behavioural patterns is built." De-institutionalization of current unsustainable norms may be hastened by political and social pressures, of which eco-groups (discussed in [Georg 1999]) are one sort. The existence of "eco-towns" undercuts "the more deterministic accounts of the pervasiveness and stability of the existing institutions" [Georg 1999].

A cross-cultural study of differences in energy use between Japanese and Norwegian householders [Wilhite 1996] provides good examples of the dependence on particular cultures and societies of the contents of the habitus. The study compared behavior and attitudes of the two populations to residential heating and cooling, lighting, and clothes and dish washing.

Traditionally, Japanese heat only the part of their dwelling in use at the time. Recently, the Japanese have substantially moved away from the traditional use of the energy-efficient *kotatsu* (a common heating unit positioned under the dining table and covered with a common comforter) as families spend less time socializing and more time in individual activities. In addition, air conditioning is increasing as a status symbol or an indicator of a socially appropriate Japanese home. Norwegians have historically heated most of their living area most of the time. The colder northern climate is one explanation, historically low energy prices another. Yet, space-heating demand was quite inelastic in the face of price increases in Norway in the 1980s. This may be because space heating in Norway and air conditioning in Japan have become "cultural energy services," i.e. they have taken on "symbolic attributes which make them integral to the culture and presentation of the home," [Wilhite 1996] and price has thus lost its incentive power, although earlier it may have played a role in solidifying the place of these energy services in their respective energy cultures.

The Japanese bathing routine is another example of a long-standing energy-intensive habit that is culturally deeply rooted in the habitus.

[43] But following a more strictly Environment-Revealing approach (Chapter 3)

Similarly, Norwegians make intensive use of (incandescent) lighting and energy-intensive space heating to maintain "coziness" in the home [Wilhite 1996]. Changes in these parts of the respective habitus of Japan and Norway will require the sort of internal dynamic Wilk describes, perhaps directed by appropriate public policies. In the meantime, the best short-term conservation means is the promotion of technologies that use less energy to satisfy the same cultural energy need. For example, a new efficient bulb in Norway must provide identical light quality, and this feature must be emphasized along with the economic savings.

As another example, Japanese cultural attitudes towards hot water for cleaning are very different from Western attitudes. In Norway, clothes are washed at 50 or 60° C. Although cleanliness and good hygiene are highly valued in Japan, hot water use is not cognitively connected to cleanliness or hygiene. Instead, hot water is used for comfort. Thus is explained the Japanese practice of washing clothes in cold water, and often even cold used bath water, while when washing dishes, the Japanese will often let hot water run, rather than use sink basins, because they use the hot water as a means of warming the person washing the dishes. As opposed to the parts of the habitus described above, Norwegian clothes-washing habits and Japanese dish-washing habits are not so culturally important to their respective cultures. These could be more easily changed through reeducation [Wilhite 1996] – e.g. shattering the Norwegian hot water hygiene myth or teaching the Japanese not to keep hot dish water running, perhaps both by presenting the other culture's successful practice as a counter-example.

Anthropological/sociological treatment of needs, wants, and restraint

Reversing the social transformation of luxuries to necessities involves inducing reconsideration of practices which are taken for granted by bringing them into the realm of discourse and questioning where "needs" can be reconverted into "wants." This kind of targeting holds particular promise for "discretionary" energy consumption that is driving a significant portion of the more recent consumption explosion, but it is probably more easily applied before the consumption behavior has undergone a long period of cultural naturalization.

3.4 Economic treatments

3.4.1 Conventional neo-classical treatment of consumption

Consumption of goods and services is the lifeblood of modern economies and the accepted primary source of welfare. Economics does not consider consumption problematic. Although the idealized consumer in neoclassical microeconomic theory aims to maximize his or her "utility (a measure of well-being)," for various reasons economics has taken as its de facto aim the maximization of the individual's consumption rather than his or her well-being. Economics posits no natural sufficiency point for the consumer: while the graph of an individual's consumption of a single commodity displays a finite "bliss" inflection point, there is no built-in limit to the type and number of commodities he or she may desire, and greater consumption of an ever-greater variety is implicitly the aim [Goodwin 1997a]. Insatiability is axiomatic among mainstream economists and, translated to the macroeconomic sphere, this premise underlies the belief in the desirability and inevitability of open-ended economic growth [Princen 1997].

Mainstream economics does recognize that specific patterns of consumption or excessive consumption of certain resources may have undesirable economic and non-economic side effects. Environmental economic consequences of consumption include, indirectly, the impacts of waste products on economic production, distribution, and consumption, and the effects of production processes on natural resources [Stern 1997b]. Localized "over-consumption" vis-à-vis these effects is possible for specific products in certain cases, especially when price signals are distorted by subsidies or, generally, cost externalization. This is commonly dealt with by applying a unit tax equal to the marginal external cost inflicted. For the individual and in the aggregate, however, over-consumption does not seem theoretically conceivable.

Although problematic consumption and over-consumption have been addressed largely outside of the discipline, some economists have, in response to criticism, taken up the defense. Others, some of whom are presented next, have departed from the mainstream to consider parts or the whole of the system as a problem, for ecological, social, or other reasons [Goodwin 1997b]. The evidence of growing, pervasive environmental damage from economic activity connected to consumption and production has caused several prominent economists to voice alternative views. Galbraith, for example, asserts that the consumer society is for the fortunate few; its extension to all the poor of the less developed world (at an American level of resource consumption) would endanger life on earth [Galbraith 1997].

3.4.2 De-coupling consumption and welfare

The issue of the link between consumption and welfare, or the lack of it, is an appropriate subject with which to bridge psychological/social treatments and economic ones, since it involves needs and wants and positional consumption. The de-coupling attempt has already been cited as an approach that questions the benefits, rather than emphasizes the costs, of high levels of consumption. It is also one that can be based mostly on economic arguments.

Psychological studies in Great Britain and the US have shown that much of the historically recent increases in material consumption in these countries is connected to largely unsuccessful attempts to meet non-material needs, as described earlier [Jackson 1999]. Other studies on positional consumption point up the large relative aspect of personal welfare. Increasing consumption has long been enshrined in economic and government policy as the major tool for personal and aggregate welfare enhancement. Yet, each of these kinds of studies, one on the individual level and the other in the aggregate, challenges the underlying assumption that welfare is necessarily served by increasing consumption in the developed world. This section examines these critiques, evidence for the divergence of national welfare and consumption, and a few counter-arguments.

The lack of a straightforward equivalence between increased consumption and improved individual welfare, contrary to economics' operationally imposed objective, has been described above. "The economic success of the prevailing system has derived from its ability to expand and create new markets for new material products. But the relationship between these new, mass-produced material products and the satisfaction of underlying human needs is no longer clear; for the very reason that the remaining needs are not really material needs" [Jackson 1999]. A large portion of individual expenditure has been found to be an attempt, through mostly resource-intensive consumption, to satisfy non-material needs (e.g. affection, participation, etc.), but these needs are largely not thus met, and therefore such consumption may provide only a pseudo-satisfaction of non-material needs and at worst may inhibit their satisfaction [Jackson 1999].

When transferred to macro-policy, such critiques question whether traditional forms of economic development that rely on constantly increasing material consumption can maintain or further long-term aggregate welfare. The message of the strongest critiques is that far from furthering national welfare, existing patterns of consumption in most developed nations, and the patterns that are being exported to less developed nations, fundamentally threaten it. Reduction of material profligacy in the aggregate can directly improve human welfare, quite apart from environmental and global equity arguments in favor of reduction. "Revisioning the way we satisfy our non-material needs is not the bitter pill of eco-fascism; it is the most obvious avenue for renewing human development" [Jackson 1999].

The majority of environmental and ecological economists remain closer to the mainstream. Many support the conventional economic aim of maximizing consumption but insist that this consumption account for environmental costs in its pricing, as in green accounting. Since the goal is still consumption maximization, and conventional economic aims remain unquestioned, the consumerism-welfare disconnect is not addressed [Lintott 1998]. Scale effects, which defeat attempts at marginal amelioration of the environmental effects of unit consumption, also remain. Dematerialization approaches can be similarly criticized.

Perhaps the least ideological argument against a consumption-based national welfare policy is that if consumption has a large positional component, as it indeed seems to have, then it is impossible to raise aggregate welfare by increasing consumption of this sort. Once absolute (basic) needs are met, individuals' increase in consumption often has more of a positional element. The empirically noted rise of expectations and standards of happiness with rising incomes supports the theory of the relative basis of welfare [Easterlin 1995]. Easterlin's economic model predicts that increases in aggregate income do not increase overall happiness since the relative disparities remain, and happiness seems to depend largely on perceptions of relative conditions of betterment [Easterlin 1995]. Others point out that for society as a whole, positional consumption is a zero-sum game: An increase in total welfare from positional consumption is "definitionally impossible." In fact, one may generalize that any type of consumption that increases individuals' welfare only relative to others (i.e. psychological welfare, image) cannot, in its expansion in the aggregate, provide an aggregate increase in welfare [Lintott 1998].

Empirical evidence seems to support such arguments about the absence of a continuing link between consumption and welfare in developed countries. Careful polling and measurements in many studies of populations in the US, Europe, and Japan over the past few decades confirm that people's sense of happiness does not increase linearly with increasing per capita income [Goodwin 1997b], [Easterlin 1995]. Despite greatly increased spending over the post-WWII decades, Americans do not report significant and corresponding increases in happiness [Durning 1991]. "There is little evidence for substantial increase in needs-satisfaction in ... categories {of consumption} and considerable literature suggesting modern society is increasingly suffering from varying degrees of poverty in relation to them" [Jackson 1999].

Application of Daly and Cobb's (1994) Index of Sustainable Economic Welfare to several developed countries and at least one less developed country all show a leveling-off or decline in aggregate welfare by the mid-1970s or early 1980s and thus a divergence from continued GNP growth from that point forward [Jackson 1999]. As an explanation, Max-Neef posits a threshold hypothesis whereby economic growth, initially an unmitigated good for the national economy, eventually incurs enough environmental and social costs to reduce net human welfare. The environmental bads experienced consist of depletion of natural resources and infliction of various other kinds of damage on the natural environment, largely because economic growth has been, and continues to be, so closely coupled with growth in material and energy throughput [Jackson 1999]. There is tentative empirical evidence for the existence of an inflection point in graphs of national welfare versus aggregate consumption at much lower levels of consumption than those of North America and Western Europe [Lintott 1998].

As counter-evidence to the attempt to de-couple consumption from welfare, several studies have shown that on average people in affluent countries are happier than those in poorer countries. Others have found a small but consistent correlation between income and happiness within developed countries [Gatersleben 1998]. Evidence aside, the great difficulty in getting political support for measures to reduce (growth in) consumption must be confronted, particularly when profit is linked largely to sales volume. Still more difficult are the social and psychological complexities of the relationship of consumption to the satisfaction of needs. Even granted that the neoclassical view is too simplistic, the critical view, arguing that beyond a certain point an increase in aggregate consumption does not increase overall welfare, may underestimate the importance of consumption. Consumption is "woven into everyday life, the activities that are decisive for the quality of life and the images of the good life, so consumption is difficult to isolate as something that can be reduced without diminishing the quality of life" [Røpke 1999]. The degree to which people rely on household goods for their perception of quality of life must certainly complicate simple attempts to separate consumption from perceptions of welfare.

De-coupling consumption and welfare

Some studies challenge the conventional assumption that individual and aggregate welfare is necessarily served by increasing consumption in the developed world. The strongest critiques assert that reduction of material profligacy in the aggregate can directly improve human welfare, even apart from environmental and global equity arguments in favor of reduction. If consumption has a large positional component, then it is impossible to raise aggregate welfare by increasing consumption of this sort: Positional consumption is a zero-sum game. There are theories and tentative empirical evidence that national welfare leveled off or declined at much lower levels of aggregate consumption than current levels in North America and Western Europe.

Daly and Cobb's and Max-Neef's graphs indicate suggestively a discrete point of national aggregate *over*consumption. Is there a sufficiency point, an optimum level of consumption beyond which well-being stays constant or decreases? How can it be defined? Although inimical to the dominant cultural regime, there have likely been, and could again be, societies that operate according to the rule of sufficiency [Goodwin 1997b]. The concept of sufficiency in a socio-ecological context and the associated concept of restraint are taken up next.

3.4.3 Restraint and sufficiency

Restraint is a resource management concept with important implications for socio-economic institutions and policy. Sufficiency, its macro-level analog, has obvious economic relevance. Princen's notion of restraint is grounded in both ecology and social science; it has appropriately been described in *Ecological Economics*, among other journals. Princen's restraint and sufficiency are discussed here, while further micro and macro-economic implications are explored in the following subsections.

Princen is concerned with how to translate global ecological constraints into signals and mechanisms apparent at a micro-level in everyday life. In principle, social limits to consumption may be imposed by material and energy availability, the lack of need or demand, external constraints like government regulations, and constraints internal to individuals or groups like religious or cultural proscriptions. The first three mechanisms are inadequate to establish sufficiency, while the fourth, internal constraints or restraint, has the greatest potential [Princen 1998]. Wilk's dynamic cycle of needs and wants is an example of such an internal societal mechanism that can both constrict and enlarge the notion of "enough." Another potential mechanism is the environmental Kutznets curve, but in several places we have raised counter-arguments to the theory that people or nations adopt environmentally friendlier (values and) lifestyles as their wealth or income increases.

Princen emphasizes that individual and social restraint is not "evolutionarily novel behavior." Human beings have long had to contend with, and adapt to, resource limits. Restraint has been de-emphasized by modern society that defines progress overwhelmingly in terms of material consumption [Princen 1997b]. While environmental restraint is by no means innate in human beings, successful civilizations have cultivated it in order to manage their common resources within natural constraints [Ponting 1991], probably often by assertion of ownership, either individual or communal [Ridley as cited in Princen 1998]. To instill restraint in the use of a resource, it is necessary and sufficient – absent education, coercion, and incentives –

to receive feedback from resource consumption as to the consumption's impact on the resource stock. Negative feedback concerning the security or stability of the resource illustrates for the consumer the risks of his behavior to his economic security or survival.

Restraint is the "deliberate reduction in immediate consumption for material benefit in return for nonmaterial benefit." Technically, it is an individual's consuming less than possible at present. Examples of nonmaterial benefits are group acceptance, reduced uncertainty, improved reputation, and ethical fulfillment. A simple example of restraint is desisting from eating seed and waiting until the planted seed yields crops. Restraint is not altruism. Altruism cannot be depended upon as a societal mechanism, particularly when large-scale social change is necessary. Conservation, of energy or wildlife for example, comes closest to the concept of restraint. "Yet restraint is done not to help alleviate a known problem but for reasons of self-management which arise internally and systemically" [Princen 1997b].

Because exercise of restraint yields non-material benefits, it is largely out of the scope of normal economic reasoning. Restraint occurs at all levels of income and material wealth. This also contests the implication of the environmental Kutznets curve that a certain threshold of aggregate material wealth must be surpassed before people can concentrate on "conserving." Whenever people choose a non-material pursuit (e.g. community volunteering) over a resource-consumptive one, they are, at least temporarily, exercising restraint [Princen 1997b].

Princen places restraint and substitution at either end of a spectrum. Substitution puts the consumption of one thing in place of that of another, with ambiguous effects on the overall resource base. Restraint reduces consumption in return for an increase in non-material benefits of some kind, with unambiguously lower resource consumption [Princen 1997b]. This characterization of the difference between substitution and restraint on a micro-level goes to the heart of the difference between macro-consumption policies that favor changes in consumption patterns and those which favor reductions in consumption levels, a difference discussed at length at the beginning of this chapter.

The practice of restraint among social groups determines and maintains points of sufficiency. In general, sufficiency can be defined with respect to human welfare and happiness or with respect to environmental sustainability [Durning 1991]. Defined with respect to the environment, the sufficiency point is that time and level at which consumption of a resource stops growing and levels off or declines. The collection of sufficiency points for all resources feeds into the notion of an aggregate material (and energy) sufficiency point for society as a whole. Such a quantity is Daly's elusive aggregate throughput limit. In principle, "if sufficiency is reached on all critical resources and the associated level is regenerative, a sustainable economy is achieved" [Princen 1997b].

While many concerned with overconsumption expect it to be remedied either by limits imposed by governments or naturally (and potentially catastrophically) by an encounter with biophysical limits, the notion of social restraint suggests other policy avenues to pursue along with the traditional economic or regulatory measures [Princen 1997b]. Wilk agrees that as a policy tool, supporting natural constraints or social inhibitions for consumption may be more fruitful than incentives, addition of new constraints, or changing desires or perceptions of need [Wilk 1999]. Policy makers should look for ways to foster restraint's favoring non-material consumption over material consumption. This requires research into the question of which conditions lead restraint to become dominant and not submerged, as it is currently in the prevailing neo-liberal regime of technologically-driven economic growth [Princen 1997b]. These conditions are explored in the remaining economic subsections. Specifically, Princen's theory of work-induced restraint and the structural economic changes necessary for reintroducing restraint (e.g. lessons from common property research) are discussed below in, respectively, the subsections "Breaking the work-and-spend cycle" and "Structural economic factors."

Restraint and Sufficiency

Social limits to consumption may best be imposed by constraints internal to individuals or groups. To instill restraint in the use of a resource, it is necessary and sufficient to receive negative feedback from resource consumption as to the consumption's impact on the resource stock. The exercise of restraint yields non-material benefits and under the right circumstances occurs at all levels of income and material wealth. The practice of restraint among social groups determines points of aggregate sufficiency for resources, and the collection of sufficiency points for all resources theoretically defines aggregate material and energy sufficiency for society as a whole.

3.4.4 Breaking the work-and-spend cycle

Princen has argued that there are natural restraints inherent in the traditional organization of work, restraints in both personal labor input and in the scale of output produced [Princen 1998]. The labor-limiting restraint mechanism operates when workers limit their needs and desires in order to reduce the amount they need to work in the economy to satisfy those desires. The mechanism is linked to the oft-dismissed notion of the backward-bending supply curve, according to which at some point on a scale of continuously increasing wages or earnings, workers cut back the amount they work, *if they are given the opportunity.* Accordingly, there is natural sufficiency in laboring: It is universally true that when people can control their work conditions and quantity, they adjust and limit them; they "minimize externally directed and compensated activities and maximize internally satisfying productive activities."

The key condition is italicized: Workers must be able to limit their hours, but this is generally not the case. Labor efficiency can be improved in two ways: More output for a given input or less input for same output. The first is almost always chosen and is the hallmark of the growth economy. More things (and more things to tax) have been popularly and politically more appealing than greater amounts of leisure, so Princen speculates. Thus, while the quality of work in industrialized countries has often deteriorated with the mass production of goods, and the work ethic and status of work have also diminished, work hours have not consistently declined [Kiron 1997].

Working from a different economic perspective, Schor has reached the same conclusion as Princen regarding the need to empower workers to limit their work time [Schor 1991, 1995, 1998]. For some time in the US, Japan, and many European countries, employers have chosen to relay the gains of higher productivity mostly as higher wages instead of reduced hours. Employers stipulate work schedules for most of their employees, and the workers conform. Workers use their increased monetary compensation to increase their consumption (or, by investing, their future consumption) and with time, they may become accustomed to this higher level of expenditure (or addicted to it), such that they would be unwilling to reduce spending in return for less work and more leisure, even if it were offered to them anew. Thus, their preferences adapt and reinforce the environmentally undesirable systemic arrangement. Schor calls the result the "work-and-spend" cycle, and for social and psychological reasons as much as for environmental ones, she urges change to break the cycle.

Despite their conditioned desire for increased consumption, polled workers in all segments of the workforce across the US and Europe evince a desire for less work and more leisure time. Many show a willingness to trade consumption for more leisure [Schor 1995]. For the successful but long-hour working, harried employee, leisure time is indeed one of the most coveted items that money cannot buy. Although some large companies such as Volkswagen in Germany have instituted flexibly-chosen part-time work for a large portion of their work-force (which can therefore be larger than otherwise), most companies frown upon part-time work. Long hours and overtime are often unavoidable or expected as a sign of worker devotion in higher-level work. The trend towards corporate downsizing has resulted in even longer hours for employees retained. And there are signs that the length of the European workweek is again increasing.

Many other critics, including Durning (1991), also view policies that encourage consumption, particularly materially intensive consumption, at the expense of leisure as among those most in need of change. Of course, many popular forms of leisure activities are now highly material and energy intensive; for this, different sorts of policies and signals are necessary. Schor summarizes the call for change in the following: "Commitment to an expanding *material* standard of living for everyone – or what Galbraith has called the "vested interest in output" – entails our continuing confinement in the 'squirrel cage' of work and holds the potential for ecological disaster. Or, we can ... realize the promise of free time which lies before us" [Schor 1991, italics added].

In addition to time limits on a personal level, Princen suggests that scale limits on a collective level are natural to the self-management of work and can either be encouraged or inhibited by policies and institutional arrangements. Princen's general concept of restraint in resource use just presented requires ecological feedback to the exploiter on the state of the resource. In short, this implies that conditions favorable to restrained use include "exclusive, small group use of an essential and well-defined resource," and probably generalist rather than specialist control [Princen 1998]. The modern global economy is characterized by precisely the opposite conditions – a high degree of specialization and enormous separation of the use of resources from the consequences, dispersed through innumerable levels of ownership and production. These types of structural features appear to be among the most fundamental and challenging obstacles to sustainable consumption.

Break the work-and-spend cycle

There seem to be natural restraints inherent in the traditional organization of work, restraints in both personal labor input and in the scale of output produced. The labor-limiting restraint mechanism operates by workers' limiting their needs and desires in order to reduce the amount they need to work in the economy to satisfy these desires. But employees must be given the opportunity to limit their work time. Instead, many workers are caught in a cycle of "work-and-spend," although polls shows a desire for less work and more leisure time. Institutional changes should permit greater (energy-extensive) leisure instead of longer hours and concomitant higher consumption.

3.4.5 Structural economic reforms[44]

Basic economic driving forces for consumption include persistent productivity increases, driven by competition, some forms of which are intended to attract customers; product innovation, which has become even more important competitively than process innovation; functional obsolescence; product diversification and specialization; advertising, especially the general selling of the idea that all of life's problems have a solution in a product or commodity; and the growth of easy credit for purchasing [Røpke 1999]. Parts of the logic of the consumer economy are fundamentally at odds with sustainability of the strong variety and conflict potentially with the weak version as well.[45] Even according to weak sustainability, the consumption level should be lower than that permissible in the absence of significant environmental degradation, since more man-made capital must be added to the capital stock to compensate for declining natural capital [Lintott 1998]. Consumerist logic dictates the consumption of an ever-greater quantity and variety of goods ("macro-insatiability"). The vested interests in maintaining this growth are enormous, and the growth momentum is still greater in an integrated global economy. If scale is a concern, then government measures amount to "add-ons" that "at best ameliorate degradative trends associated with an infinitely expansionist

[44] A complete treatment is not possible here and would require its own study of at least commensurate size. For a recent example of such a study, see [Hawken 1999].

[45] Many economists justify open-ended consumption of natural resources, and even their degradation, if done at least in a weakly sustainable fashion. Weak sustainability is the substitution for consumed or degraded environmental capital of a supposedly equivalent amount of man-made capital. Weak sustainability is achieved if the natural degradation rate is matched (or exceeded) by the rate of creation of man-made capital. In strong sustainability, natural resource consumption is compensated by an equivalent investment in natural resources of a comparable type.

From a biogeophysical perspective, weak sustainability is inadequate for several reasons. First, substituting improved technology for environmental services is problematic: it entails additional costs and may produce new environmental impacts. These effects may be greater than the benefits of the technological substitutes. Second, technology cannot be counted on adequately to replace any degraded natural service in time and with a matching quality and scale [Holdren 1995]. Further, the First Law of Thermodynamics means that natural capital and man-made capital are not independent: natural capital is often needed to produce man-made capital; only natural capital can realistically perform life support functions of the global biosphere; substitutability between the two from the perspective of economic theory is not so clear; natural capital can exhibit irreversibilities; and huge uncertainties about impacts of the loss of natural capital make tradeoffs between natural and man-made capital very difficult to conduct [Munasinghe 1995a].

economy" [Princen 1998]. Such measures are particularly inadequate for controlling overconsumption whose costs show up much later or far from the locus of activity.

Structural factors and political decisions in the industrialized economies significantly limit the element of choice in consumption; alternative choices are bounded by the social-economic-technical framework [Røpke 1999]. Consumers are thus limited in their latitude to respond to appeals for change. "Understanding how our present choices are self-defeating is a crucial step in the process of change, but so too is understanding how the social and political context makes such self-defeating choices seem almost inevitable..." [Wachtel 1983]. Sustainability seems to require specific redesign of elements in the consumption and production economy, rather than evolutionary change [Uiterkamp 1998]. Whether the pressure for change should come from the top or the grassroots citizen-consumer level is considered shortly in the chapter conclusion.

In keeping with his emphasis on restraint, Princen finds that the resolution could come from changes in the political economy to restore an "ecological" type of constraint [Princen 1997a, b]. Decision makers in the economy must be in a position to receive and react to ecological feedback (analogous to individual workers' adjustment of their kinds and levels of work in response to direct biophysical feedback) [Princen 1998]. Thus, establishment of property rights is an effective policy intervention since certain forms may allow for direct feedback from consumption. Self-management induces restraint to increase long-term personal economic security [Princen 1997b]. In modern economies, however, decision makers function at a great distance from the loci of resource extraction, use, deposition, and direct ecological consequence [Princen 1998, 2002]. Business strategy and government policy are oriented around the search for market frontiers whose exploitation generates large indirect costs. Separating consumption and production decisions across large distances of geography, culture, and agencies ("distancing") generates large gaps between benefits and ultimate full costs, costs which are also further systematically obscured ("shading"). The distancing and shading of commerce must be reduced to reestablish the possibility of restraint-inducing feedback [Princen 1997a, 2001, 2002].

Cogoy's recommendations for structural change follow from his view of consumption as a process having both market and non-market inputs [Cogoy 1995]. The economy strongly favors (specialized) market inputs at the expense of non-market ones: Products that can be produced industrially consistently fall in price compared to products that cannot be so produced. This happens by increasing the division of labor and substituting capital for labor, both of which are enabled by a system that permits externalization of the social and environmental costs involved [Røpke 1999]. The position of the border between market and non-market production and consumption can be influenced institutionally. Environmental protection may require shifting the border more towards non-economic consumption, even though this shift would imply changes in the existing power distribution [Cogoy 1995].

The shiftable border between market and household consumption decisions is well illustrated by both household energy conservation and home power production or co-provision. Household energy conservation returns some of the skill and planning in consumption from market enterprises to the householder. Cogoy assumes that failures in household conservation efforts have much to do with an inadequate institutional definition of the market/non-market border [Cogoy 1995]. In the case of the American home power movement in which participants generate their own electrical power, the shift towards private production and control over consumption has in many cases reduced environmental damage and increased personal and community control. Home power users seem to internalize almost all of the environmental and social effects of their energy choices by the generation and highly controlled use of their own power [Tatum 1995, Princen 2002]. Similarly, successful application of environmental taxes, in transportation for example, requires that the social structure provide flexibility and facilitate innovative consumer responses that may lie outside conventional market choices [Cogoy 1995].

The need for full-cost pricing and the removal of enormous government subsidies for resource extraction, depletion, transport, and related activities is a common prescription of eco-efficiency. Less commonly heard, but perhaps just as important, are fundamental points like those just discussed, as well as older critiques of the consumer economy like planned obsolescence [Charkiewicz 1999]. Grappling with the scale problem requires not only the eco-efficiency goal of increased efficiency of resource use but also determining which goods and services are dispensable, redundant or largely discretionary [Daly 1996a]. Even further, sustainable consumption may require breaking out of the near-universally accepted, linked imperatives for the joint growth in consumption, production, GNP and stock market indices, and employment. This may require reforming not only the public finance system – including shifting the tax base from employment to natural resources and eliminating perverse resource subsidies – but also the international monetary and financial system, reducing the high levels of debt that compel enormous production and consumption to service them [New

Consumers Conference 1999],[46] and even changing the status of money as a primary external goal of economic activity [Jaeger 1994].

It seems illusory to think such significant changes will be wrought voluntarily. Yet environmental realities are finally being registered in the prices of certain resources as they are severely degraded or depleted: This may force producers to reduce the associated resource throughput per unit of production. If this happens on a scale sufficient to lower aggregate resource throughput, monetary policy and other economic macro policies and institutions will have to change accordingly [Goodwin 1997c]. Charting directions for such changes is an important area of economic research.

Structural economic reforms

Various structural and institutional reforms to the economy have been proposed to facilitate sustainable consumption, including restoring feedback to decision makers by establishing property rights and reducing systematic obscuring and distancing of costs; shifting the market/non-market border in favor of non-market consumption; full-cost pricing, elimination of resource subsidies, combating systemic planned obsolescence, shifting taxes from work to resource consumption, and, fundamentally, reforming the monetary and financial system to control the growth of money and debt.

4. CONCLUSION: POLICY SYNTHESES AND POLITICAL IMPLICATIONS

4.1 Altering consumption: top-down or bottom-up?

This section considers a question important to the synthesis of the various approaches and critiques in interventions for sustainable consumption: namely, which of the two broad types of incentives to alter consumption,

[46] Whiston identifies high debt levels and the structure of the world economy as a central reason for the South's severe environmental problems in the latter part of the 20th century. It was in the service of domestic and Northern economies that many of the Southern states were largely geared around disastrously ecologically damaging activities such as single-crop production or mining. More specifically, Northern countries have been directly at fault for purchasing these mineral or agricultural products at prices that do not account for the huge environmental and social costs they inflict in the countries of origin; and the North has been indirectly culpable by setting terms of trade that force the South to liquidate natural resources in order to pay for products of the North – products that they are unable to produce themselves, partly by forced specialization, and that they covet, partly on the strength of the Northern example and marketing [Whiston 1990].

internal "bottom-up" or external "top down," are the most motivating and most durable; and the related question of whether individual change in behavior alone is effective or whether societal change is necessary. In the literature, the responses to these questions depend again on the authors' disciplinary orientation. Most consumption critics see the need for "changes on a society-wide level, to enable and support individual lifestyle changes, away from the behaviors associated with a consumer society." They often assume a large role for government in this, and they assume that "in a democracy, such government action may – or should – require a shift in relative values within the population as a whole" [Goodwin 1997c]. Policy makers typically prefer a "top-down" approach of informing and educating citizens about the necessity of making changes [Georg 1999]. This task is vitally important, as described in the final chapter section. Yet, isolated attempts to prompt such changes often fail, in part because consumption is multiply determined, as has been amply illustrated. Effective policies must simultaneously target several barriers at once [Stern 1997a]. But which sort of barriers on which levels are most important?

The free-rider (social trap) problem is said to stymie individual efforts at behavioral change. Apart from the exceptional few, most people will not act unless everyone acts, while those who act fail to make a difference overall [Meijnders 1998]. Georg (1999) claims that free-rider problems can be overcome if new norms of behavior are established and institutionalized. Treadmill-of-productionists doubt that even new norms can overcome deeply entrenched structural problems with the global economy and North-South relations. Advocates of "voluntary simplicity" lifestyles believe change from the bottom-up is possible. Yet, unanswered is the question of how large such micro and local movements would need to be to affect the macro-economy. Working from the bottom-up, Eco-team programs[47] assume that 15-20% of the population is enough to constitute a critical mass to sustain environmentally friendly changes in society. However, the Eco-team group-session approach is clearly not suitable for such a large portion of the population [van den Burg 2001]. Also at issue is the negative effect the growth of such movements would have on the macro-economy of the majority who would not have changed from the consumerist paradigm [Goodwin 1997c]. Could such change be accomplished piecemeal, and, in a globally interconnected economy, could it be attempted by individual

[47] The Eco-team meetings used by environmental groups in many countries allow people to track their resource consumption and waste generation as well as compare themselves and exchange conservation tips with members of their peer groups (see section 1.3 in Chapter 5).

countries without systemic ripple effects and "punishment" of the diverging pioneers?

Structural limitations on choice constrain individuals' latitude to change their lifestyles. Why do most writers on sustainability, with their oft-repeated phrase *lifestyle changes,* seem to emphasize individual and group practices over social consumerist norms, institutions, or economic barriers? Basic political pragmatism is certainly one reason. Yet, while lifestyles in the sense of social practices are close to the root of the issue, the systems that facilitate and encourage destructive lifestyles need to be changed; individual alternative practices, absent top-down incentives for new socio-technical pathways, are likely to remain anomalous against overwhelming trends. There are some promising examples of fairly minor adjustments in institutional frameworks that have engendered disproportionately large, favorable responses [Tatum 1995] and brought about greater positive change than anticipated. In general, we conclude that change in a society's consumption trajectory requires combinations of individual, social, institutional, and technological change. Political or economic incentives in the absence of modification of personal or social practices (or values) tend to be superficial and ineffective; but individual or localized action alone such as voluntary simplicity is unlikely to work societal changes either [Durning 1991].

True to the aim of Wilk and Whiston, establishment of new (or renewal of old) social norms in favor of restraint shows the greater promise of long-term, psychological, and institutional durability, in contrast to application of economic instruments without such a change in norms. However, "absent crisis, they may also be the most difficult to implement in the face of a dominant belief system – namely, technologically driven economic growth" [Princen 1997b]. The potential for information to modify consumption behavior and/or enhance environmental and socio-political awareness, and the most effective form and means of communicating it, are discussed next in the upcoming chapter. In the context of this section's themes, however, Princen's caveats bear mentioning now. Much environmental education is premised on a perceived lack of information for actors in society. However, many now believe that greater information and public awareness alone will be insufficient to control the pace and scale of environmental degradation. Changing attitudes as an approach to instilling restraint is difficult and would not occur fast enough or on a sufficiently wide scale to stem the growing damage [Princen 1997b]. In practice, it has been found that information and education initiatives alone, without further incentives, have had low levels of success [More with Less 1995]. Such information, learning, and economic and behavioral incentives run counter to dominant social forces. Instead, Princen favors focusing on and changing conditions of ecological feedback, institutional arrangements and economic security [Princen 1997b]. While some try to use information or persuasion to convince consumers (or businesses and governments) of the threat that overconsumption in the aggregate poses to their health or security, Princen argues for changes in economic and institutional arrangements that bring to individuals' attention the threats micro-misconsumption of their own resources pose to their individual (or corporate) security. But how does one muster political support for these types of change? Whence comes the political stimulus for the necessary restraint-enabling institutional changes, when only these changes would bring home the reality of personal and organizational threat in time? Like most chicken-and-egg problems, this one has no easy answer. The use of innovative information and educational tools is one of this book's main contributions towards its solution.

Altering consumption: Top-down or bottom-up?

Change in a society's consumption trajectory requires combinations of individual, social, institutional, and technological change. Political or economic incentives in the absence of modification of personal or social practices are superficial and ineffective; but individual or localized action such as voluntary simplicity alone is unlikely to work societal changes either.

4.2 Political implications and alternatives

Røpke and others conclude that a wide spectrum of measures and the involvement of many would be required to effect a meaningful alteration of the North's consumption trajectory. This involvement presupposes a degree of consensus on the need for, or the benefits of, altering or curtailing resource consumption: The present consensus is in favor of continued consumption growth. The fundamental economic and institutional means of altering this trajectory may involve changed patterns of trade that reduce the transfer of resources from the South and the systemic internalization of social and environmental costs of industrial and consumption growth [Røpke 1999]. But this is a long-term ambition. The immediate aim is to garner public and political recognition and support of the need for change in the direction of sustainable consumption and production, in a time frame that responds to the exigencies of the environmental impacts of consumerism [Symposium Sustainable Consumption 1994, Durning 1991, Whiston 1990, 1999]. Preferences may change on time scales of a human generation or longer: This alone rules out the default course of "leaving change to the market" [Stern 1997c]. Whiston is pessimistic that the public can be persuaded to support the scale of long-term changes that, in his opinion, are necessary for meaningful effect [Whiston 1990, 2000].

For Whiston, the two axes of a solution to global environmental problems, North-South relations and the social contract between the state and its citizens, must ultimately be compatible, and North-South relations are disrupted by the demands of Northern populaces for excessive material wealth. Pricing to reflect the real cost of goods and services would reduce the *material* standard of living as defined by the total freedom to acquire goods, travel at will, support functional obsolescence, and the like. Understandably, governments, businesses, and consumers do not embrace such costs, which could require fundamental changes in society, economy, and politics. Thus our "immediate political task" is Whiston's ultimate question for Western democracies: How can their governments persuade electorates of the "'limits to legitimate delivery'; and more importantly how [can] electorates be encouraged to elect such governments?" But to date, this has not happened, and historically, given a choice, radical change has never been embraced by electorates [Whiston 1990]. Instead, governments and electorates in the US, for example, have often pursued "politics of default" in recent years with respect to energy supply and other environmentally critical policies, policies characterized by collective momentum and popular nonparticipation. "Nonparticipation is evident in virtually all consumer behavior... in our generalized acquiescence in accustomed patterns of behavior..." Consumers also tend to practice "'technological somnambulance'; In particular, we do not insist that policy makers and technical people seriously explore with us more than one set of socio-technical alternatives" [Tatum 1995]. Yet, as Lovins pointed out in the 1970s, there are no alternatives, including "default" paths, that do not involve untried change on a large scale[48] [Lovins 1977]. Business as usual affects our relationship with the natural world and with others in society, often quite negatively [Tatum 1995].

This chapter has traced driving forces of end-use consumption and critiques of consumerism; although it has proffered suggestions for types of changes in policies and emphasis, it has not been ambitious enough to suggest full-scale alternatives. It has been easier to identify in theory that which is unsustainable than to describe in particular what would be sustainable. In conclusion, to compensate, we offer hints of others' visions

[48] "Population growth, conventional resource depletion, expanding degradation of the environment, and the inevitable implications of carrying on 'business as usual'... will bring fundamental and far-reaching change whether we like it or not. We do not have the option of preserving our way of life as it now stands" [Lovins 1977].

of alternatives to consumerist, material growth-dominated paradigms that are still basically consistent with capitalist market systems.[49]

There are several possible uses for the economic growth dividend thus far accrued that are compatible with an absolute reduction in per capita consumption. These include reducing poverty or great inequalities in wealth, reducing working hours (increasing leisure), and increasing expenditure on public goods and services instead of private expenditure [Goodwin 1997c]. Segal posits two general alternatives, the "graceful simplicity" society and the "creative work" society. A society of "graceful simplicity" embraces the second of the "two possible courses to affluence. Wants may be 'easily satisfied' either by producing much or desiring little" [Sahlins 1974]. In such a society, the economy is meant to satisfy basic needs for health and security; and the good life is enhanced not by an increase in goods and services but by a reduction of work and an increase in leisure; the economy's performance depends on how well it meets "real material needs" and produces leisure time; work is for the purpose of generating income necessary to meet "needs"; and the measure of living standards gauges the level of leisure time and the use of leisure [Segal 1994]. Policies consistent with the work aspects of such an economy include the four-day workweek and a shorter working day, which have already been introduced in parts of Europe but have been eclipsed by growing unemployment or are under siege from the pressures of globalization.

On an individual level, the drive for simplicity inherent in such a society has found actual expression in voluntary simplicity or frugality, often accompanied by anti-consumerism. People practicing voluntary simplicity are sometimes "downshifters" who have voluntarily left high-powered careers or jobs in return for more leisure time, control, and/or meaningful work. They may otherwise be motivated by a concern for the environment, a frugal attitude towards money, unemployment, or large consumer debt, or they may be members of anti-consumerist religious groups [Schor 1995, 1998]. In general, they see reducing consumption as a way to improve their quality of life. Some look for ways to give up the stress connected to pursuing higher consumption while still "achiev[ing] the essential social contracts without it" [Røpke 1999].

The obverse "creative work" alternative economy increases well-being by generating satisfying work; the economy is measured by success in providing this work; and the standard of living depends on its quality. Both the graceful simplicity and creative work alternatives do not necessarily

[49] The author has no illusions about the atrocious historical environmental record of industrial communist states.

imply increasing output and throughput. However, technological innovation and increased productivity are important in both. Both alternatives require controlling the costs of basic needs as well as controlling the "continual expansion in the definition of those needs" [Segal 1994]. The creative work alternative would also require radical change to effect; it seems highly unlikely in an increasingly densely populated world of rampant unemployment. In addition, there are no current examples to point to.

Late 19th century America, however, does furnish an example of such a movement, based on a producerist rather than the expansionist-consumerist ideal that eclipsed it in the 20th century. Princen and Lasch describe it in these terms: "The producerist ideal thus embodies a profound sense of meaning in work. In fact, it rejects the producer-consumer dichotomy and instead promotes the values of identity, economic independence, and citizenship through self-directed proprietorship. Inherent in such a vision is self-discipline, striving for purpose, and limits to ever-increasing material throughput..." [Princen 1998]. Princen does not advocate a return to an economy founded on producerist principles, beset as it was by its own sort of drawbacks and problems. Still, such examples enrich the imagination of possibility and furnish elements helpful to the effort to envision and actualize the modern-day alternative of sustainable consumption.

Political implications and alternatives

An immediate aim is to garner public and political recognition and support for the need for, or benefit inherent in, a change in the direction of sustainable consumption, in a time frame that responds to the exigencies of the environmental impacts of consumerism. Examples of alternative paradigms still consistent with market democracies can be found in the literature and in the history of Western social and economic development.

Chapter 3

ENERGY, ENVIRONMENT, AND SOCIETY

Knowledge and Risk Communication

1. INTRODUCTION

This chapter looks more closely into the research question of what kinds of lay knowledge, in which communicative formats and settings, are most helpful in bringing household resource and energy consumption in line with environmentally sustainable patterns and levels.

For analytical convenience, this chapter is split into two subject areas: knowledge (message or content) (section 2) and communication (transmission or conveyance) (section 3). These areas are explored here in some depth.

In the process, the chapter develops the overarching constructs of discretionary and non- discretionary consumption introduced in Chapter 1, using theories from psychology, sociology, Science-Technology-Society (STS), and other disciplines and schools of thought. In connection with the discretion continuum, and as a guide and frame for the concepts and discussion, two research paradigms, Energy-Revealing and Social-Revealing, are introduced and their implications for the course of energy consumption research are described. As part of the first paradigm's agenda, public awareness, attitudes, and understanding of energy and related environmental issues, especially climate change, are briefly reviewed. The topics of ecological behavior and knowledge, and then related "knowledges" in their multiple dimensions and complexity, are explored. The evolution of these topics leads into a direct, in-depth treatment of the non-discretionary concept itself. The connection to ecological modernization theory and applications is explored. Then, the chapter turns to an examination of relevant public communication and risk communication topics and draws lessons for the experimental software and interviews. The chapter concludes with final comments.

2. ALTERNATIVE RESEARCH APPROACHES AND A FRAME FOR DISCUSSING KNOWLEDGE IN THE CONTEXT OF DISCRETIONARY AND NON-DISCRETIONARY ENERGY CONSUMPTION

2.1 Introduction: Two approaches and two constructs

This section examines alternative schools of energy analysis and associates the discretionary constructs with one or the other. There is a basic analytical split in research on energy use that can be called the visibility/invisibility divide or the Energy-Revealing/Social-Revealing divide. To explain this, four possible approaches to analyzing energy consumption are arrayed in a "visibility matrix" (Figure 3-1). The alternatives showcased in this chapter lie on the southwest-northeast diagonal: we will call them the *Energy-Revealing/Social-Concealing* approach and the *Energy-Concealing/Social-Revealing* approach, respectively (usually abbreviated to *Energy-Revealing* and *Social-Revealing*, respectively). This section will clarify their meanings and the contrast between them.

It is widely recognized that global climate change and many other environmental problems and risks are largely "invisible" to people in industrialized countries, as is the overall environmental impact of energy conservation actions, as well as the connections between the use of technology, energy consumption, and carbon dioxide emissions [Shove 1997]. Energy researchers respond to this invisibility in mainly two different ways, either by striving to increase energy consumption's transparency or by presuming its invisibility and trying to illuminate its social causes and drivers. The first route (Energy-Revealing) explicitly articulates energy use in the effort to optimize people's use of it. The second (Social-Revealing) takes embedded energy consumption in products, services, and systems for granted and instead tries better to define and manage services and practices that consume energy [Shove 1997]. In part, the distinction is parallel to the analytic divide between internalists and contextualists among social scientists, for example historians of technology (see [Bijker 1995]).

"Social-Revealing"

Degree of energy visibility \ *Degree of social visibility*		
	~~E-invisible/social invisible~~	**E-invisible/social visible**
	E-visible/social invisible	E-visible/social visible

"Energy-Revealing"

Cell (1,1) is a meaningless research option since it keeps both technological and social factors invisible and illuminates neither: it can be ignored. (2,1), labeled Energy-Revealing, is the mainstream economics-engineering approach. (1,2), Social-Revealing, is the alternative wider social science approach. Finally, (2,2), suggesting total visibility, could be a combination of the approaches. For example, some researchers include traditional elements of Energy-Revealing such as examining individual's conscious energy use habits, while also looking further into societal aspects (see [Lutzenhiser 1993]). For the sake of brevity and a clearer association with the discretionary/non-discretionary concepts, only (2,1) and (1,2) are examined in this chapter.

Figure 3-1. "Visibility" matrix of research approaches

Energy-Revealing has a larger analogue in the "Environment-Revealing (visibility)" approach. The following from a paper on environmental education serves as a good example: "Improved quality of life {environment} is only possible if we have an improved perception of our contribution to environmental degradation and of the steps we can take to minimise the damages we cause. This includes, among other items, the products we buy, the policies we endorse, the initiatives we support" [Filho 1998].

Tasks for the Energy-Revealing agenda include cataloging public opinion of energy use, cultivating knowledge of the effects of (conservation) actions, and examining the influence of knowledge on actions (section 2.2). Connecting new and existing knowledge and developing and applying most kinds of indicators of energy consumption also fit into this category [Shove 1997].

The Energy-Concealing/Social-Revealing approach, on the other hand, recognizes that "... energy use is shaped in complex systems that often submerge energy and other environmental concerns..." [Wilhite 2000]. For many social scientists, especially sociologists, this submersion or invisibility is not necessarily of concern; rather it is social, cultural, or socio-technical structures[50] that need revealing.[51] This is closer to how most lay end-users relate to energy use: People are interested in the services and amenities energy provides them, not energy per se, and they largely ignore details about energy except when paying the utility bills, fueling up the car, or buying a large electrical appliance. Most energy analysts and policy makers, however, are used to striving to make energy use explicit and would be challenged to work in the Energy-Concealing/Social-Revealing mode [Shove 1997].

Yet, by hiding details of the energy balance, load, and environmental insult in favor of social and institutional explication, Energy-Concealing/Social-Revealing allows one to capitalize on a useful and perhaps singular property of energy consumption: by itself, as an undifferentiated datum, energy consumption (especially when it includes embodied energies) points to consumption in general, environmental insult, economic and security risk, and even social development. In short, energy consumption can sometimes serve as an approximate surrogate for (environmental or even general) sustainability for a country or a region. This is why 2000 watts was chosen as the single target indicator for the 2000 Watt Society. This useful feature of energy consumption can get lost in the details when a technical Environment- or Energy-Revealing spotlight rather is cast.

[50] These include economic features of the sort sketched in section 3.4.5 "Structural economic reform." Specifically focusing on these features could constitute a sort of Economic-Revealing approach as a subset or variant of Social-Revealing. This approach is popular in ecological economics and related areas but is not pursued here.

[51] The broader environmental analogue here, in its cultural cast, is well represented by the following statement: "Environmental problems are produced by cultural practices, which are determined by norms, morals, world views, in other words, ideologies. With time cultural practices and underlying ideologies become naturalized, unquestioned and thus invisible. That makes deep structures of a culture very resistant to change {see Chapter 2, section 3.3.2 "Anthropological/sociological treatments"}. In order to make purposeful change cultural practices and associated power relations must first be made visible. Implicit meanings must be made explicit" [Kapyla 1998].

The division between the Revealing approaches is part of a general disciplinary divide between the conventional economics-engineering approach to energy analysis and an alternative societal (sociological, anthropological, or STS) one. Since the Energy-Revealing and Social-Revealing approaches are this chapter's leitmotifs, it is worth spending some time to explicate and contrast the wider schools of thought that serve as backdrops and from which they derive.

Conventional energy consumption research (and applied policy) is dominated by an economics-engineering paradigm, particularly the device-centered model. It is most firmly entrenched in the US, and the limited European efforts to expand upon it have not had much impact on theory and policies [Wilhite 2000]. The commonly practiced "barrier analysis" follows from the accepted rational actor economic paradigm and is useful as far as it goes. However, barrier analysis does not necessarily contribute much towards understanding social change. As a result, the nature and underlying causes of energy demand have been ignored or only superficially explored by researchers and policy makers. Contributions to energy studies from other social sciences that might have helped to illuminate these causes have declined since the mid-1980s [Wilhite 2000]. This is partly because "disciplinary incentives" to analyze energy use from a social perspective are lacking [Shove 1998].

The alternative approach is represented by a group of sociologists and anthropologists including Shove, Lutzenhiser, Wilhite, and Kempton. It is strongly informed by Bijker's social construction of technology (SCOT) theories [Bijker 1995] and the larger STS framework. This alternative school of energy analysis is admittedly iconoclastic because it sometimes calls into question social needs or purposes for energy consumption that are nearly always taken for granted in the conventional approach. In addition, it challenges the neutrality of technologies as problem solvers and suggests that they may be part of both problems and solutions [Shove 1998].

These researchers admit that their analytical perspective assumes that significant changes in aggregate energy use (and probably environmental protection) "will be predicated on a significant social transformation" [Wilhite 2000]. The alternative approach departs from the linear theory of technology according to which technologies diffuse according to predetermined, goal-oriented trajectories and instead considers the contingent character of technological development and the "possibilities and the constraints of change and choice in technology" [Bijker 1995]. Researchers of the STS (and related social shaping of technology (SST)) school see some possibilities for reversing entrenched technology choices and reopening the field to alternative paths of socio-technical development; whereas most analysts, politicians, and the public (especially in the US) expect solutions through marginal manipulation of economic or technological variables and through general technological development [Wilhite 2000]. (See section 3.1 below for a discussion of closure and reversibility.)

Another major difference in analytical emphasis is the divide between the conventional individual actor focus and the sociological, systems-organizational focus. The derivative Social-Revealing approach considers norms and social institutions as independent variables along with the usual variables of price and consumer awareness [Wilhite 2000]. Actors' interactions at various levels and their motives – competing interests, mutually imposed constraints, situation-specific factors, and varying ascribed social meanings – obfuscate the line from the individual to his impact on energy use. These wide-scale interactions, and not only the energetic or environmental dimensions of individuals' energy use, deserve illumination: These include "cultural and socio-technical embedding of energy-related practices"; co-evolution of norms, practices, and ways of life with energy-technologies; cultural norms and "shared expectations"; the role of institutions, and the historical development of infrastructures [Shove 1998].

Such issues and especially the study of demand evolution – processes of change, "meta-energy services" like comfort and convenience, and particularly the mechanisms by which energy-intensive lifestyles become widespread and then normalized – represent relatively new ground for energy analysis and go beyond the neo-classical economics approach to demand, which does not examine the history or socio-technical "construction" of preferences [Wilhite 2000].

The concept of non-discretionary use of energy and the distinction between discretionary and non (or less)-discretionary derive largely from this alternative societal mode of inquiry.[52] The economics-engineering

[52] Actually, some sociologists insist that all human technological artifacts and systems are socially constructed and that therefore, in a sense, all human energy consumption that taps into these systems is capable of being altered, i.e. discretionary. This may be true at the societal level and over suitably long time scales (cf. Figure 1-3). But at the level of the single individual or household, the intended target for energy communication in this study, much of consumption is less discretionary, especially in the short to mid-term. The processes of closure and stabilization provide technological rigidity and social solidity, as will be discussed below. On the other hand, another type of sociological reasoning runs that all individual actions are socially and culturally determined or constrained, such that all human actions are non-discretionary and no action is truly discretionary. Bijker prefers a mixture of actor and structure and of change and constancy, a combination of contingency with structural constraints [Bijker 1995]. Similarly, Spaargaren follows Giddens in asserting that domestic consumption practices (and their ecological modernization) are simultaneously actor-driven and system-imposed [Spaargaren 2000]. As described in Chapter 1, we follow this latter view by insisting on a mixture, a variable

approach (Energy-Revealing) usually tends to focus on individual actors and the potential for influencing their direct energy consumption; thus it tends either implicitly to assume most end-users' energy consumption is discretionary, or it limits its focus to the direct part that is. The alternative Social-Revealing approach permits exploration of less directly discretionary factors that work to constrain individuals' energy consumption, which is why this approach is associated with the non- or less-discretionary concept in this study (Figure 3-2).

Economics-engineering (Energy-Revealing) approach:	Focuses on discretionary energy consumption
Wider social science (Social-Revealing) approach:	Permits focus on less-discretionary factors in end-users' energy consumption

Figure 3-2. Association of the discretion constructs with the Revealing/Concealing approaches

For clarity, it should be emphasized that the terms *Energy-Revealing* and *Social-Revealing* pertain to analytical approaches; the terms *discretionary* and *less/non-discretionary*, on the other hand, apply to people's actions, choices, and understanding:

Researchers' analytical **approach:**	*Revealed* levels of *visibility* in energy use and social interaction
End-user's **actions and choices:**	*Discretionary* extent, constraints, degrees of freedom

Figure 3-3. Application of *Revealing/Concealing* and *discretionary/non-discretionary* terms

The remaining sections in this chapter relate to one or the other energy approach, or they combine elements from both. The subsections on public knowledge and opinion of energy and the environment in section 2.2.1 review issues basic to Energy-Revealing, while hinting at related fundamental social issues. The next sections on knowledge and ecological behavior, 2.2.2, begin again with central Energy-Revealing (discretionary)

---

range or continuum, of levels of discretion for the individual in his or her consumption choices.

issues but gradually move into the Energy-Concealing/Social-Revealing agenda, which is taken up in full in section 2.3 on non-discretionary factors in energy consumption. Section 3, public and risk communication, deals with the transmission of information and insights from both perspectives, with an emphasis on Social-Revealing.

2.2 Selected topics in the Energy-Revealing approach and actors' discretionary energy consumption

2.2.1 Public attitude and knowledge of the environment and energy/climate change issues

2.2.1.1 Public environmental attitudes and knowledge

To begin the sections under the Energy-Revealing/Social Concealing rubric, we take a brief look at the visibility of environmental problems for the general publics in Europe and North America.[53] We trace public opinion and knowledge of general environmental problems and then, in reverse order for the convenience of presentation, public knowledge and opinion of climate change.

Copious survey data and more limited voting and consumer data from the United States and Europe suggest that the public's environmental awareness and concern has increased dramatically over the past few decades. Acknowledgements of environmentalism have become common, especially among the youth (even while the youth's materialism has intensified, by most measures [Kaufmann-Hayoz 1999], [Kempton 1993]). Farhar detected the beginning of a convergence in the late 1980s in American public opinion and policy makers' environmental priorities (the latter represented by those in the US Environmental Protection Agency): Fifty-two percent of the public said acting to address ozone depletion and climate change was "extremely important." Seven percent identified the greenhouse effect as the most serious environmental problem, and this was the third most frequently mentioned problem. Sixty percent were "quite concerned" about global warming in 1989, 1990, and 1991 [Farhar 1994].

[53] Section B. 36.8 of Agenda 21 on "increasing public awareness" summarizes the official UN charge for environmental visibility: "There is still a considerable lack of awareness of the interrelated nature of all human activities and the environment, due to inaccurate or insufficient information ... There is a need to increase public sensitivity to environment and development problems and involvement in their solutions and foster a sense of personal environmental responsibility and greater motivation and commitment towards sustainable development." http://www.un.org/esa/sustdev/agenda21chapter36.htm

In extensive interviews Kempton found high public concern for the environmental protection of local amenities and health protection; deep-rooted concern for future generations' and especially children's environmental welfare; and the identification of environmental protection as one of the primary parental responsibilities [Kempton 1991]. Recent polls show a strong commitment to environmental protection even at the cost of reduced economic growth. However, Americans are generally quite optimistic that this cost need not be paid; in 1999, 83% asserted that a healthy (improved) environment was compatible with continued robust economic growth [Kull 2000].

Critics suggest that the public's environmental knowledge and consciousness remain superficial, symptom-oriented, and divorced from a holistic understanding of the real "environment," its condition, and societal drivers for its decline (see e.g. [Whiston 2000]). While these observers detect a strong element of lip service in Americans' expression of environmentalism, others cite anecdotal evidence suggesting people are willing to support substantive changes for the environment, although these assertions depend largely on the meaning ascribed to *change*. For example, for additional power generation, city residents have chosen more expensive, more apparently environmentally friendly power sources: Sacramento residents voted for both natural gas and efficiency as well as renewable energy rather than the pure nuclear option offered them, and Tallahassee residents chose efficiency and renewables over a clean coal option [Farhar 1994].

In Europe, environmental concern has penetrated to the mainstream European public over the decades since it was the province of "middle class radicals." Concern for the environment is now much more general, widespread, and crosscutting in politics and policy administration. Even so, a rigorous review of environmental perceptions and attitudes in five West European countries again suggests this concern is too weak to induce significant political changes in its support [Mohler 1994].

The Swiss display at least as high an increase in environmental awareness and sensitivity as Americans, if not higher. However, their general perception of conditions and their assessment of remedial prospects seem markedly more pessimistic: Two-thirds think that local and regional environments are being degraded, and most think they will worsen. Interestingly, concern, urgency, and environmental pessimism seem to be inversely correlated with age, with about 90% of the youth considering environmental protection urgent [Kaufmann-Hayoz 1999]. About half of the general population considers global environmental issues, including climate change, to be the most pressing sort [Finger 1994].

Attitudes toward and opinion of environmental (and energy) related problems are to be distinguished from the state of actual environmental knowledge. The notion of environmental or ecological knowledge will be explored in greater depth in the next section. The relationships between environmental knowledge and environmental consciousness, action, and associated social processes are particularly complex. For now, we note only that the public state of "declarative" environmental knowledge, that is factual knowledge of the workings of environmental systems, for example, is fairly limited, despite high levels of professed concern.

Measurements from Mohler show that while only one percent of the European public "are completely devoid of {environmental} scientific knowledge, so too are only about one per cent absolutely knowledgeable." Additionally, in an insight into the connections between environmental knowledge, attitudes, and values, the report found that certain measures that seem to be gauging environmental knowledge may actually be capturing environmental pessimism [Mohler 1994].

While the Swiss are highly concerned about the environment and seemingly highly aware, the public's knowledge base may actually be shallow and journalistic, since it derives mainly from the media. The Swiss public generally seeks more environmental information, albeit from these same media sources, especially concerning those issues that generate the greatest anxieties: Chemical and nuclear pollution, ozone depletion, and climate change [Finger 1994].

2.2.1.2 Public attitudes and knowledge of climate change

The public's understanding and knowledge of climate change, specifically, and their attitude towards the issue resemble their view of environmental problems in general. Here again, despite high levels of professed concern, the public's grasp of the basic scientific issues involved has lagged behind, although there are recent signs of change.

Popular notions of human-induced climate change have surprisingly old historical precedents. Americans and Swiss thus seem culturally primed to believe in the possibility of anthropogenic climate change [Kempton 1991], [Pfister 1998]. While a readiness to embrace the concept is in place, the public's understanding of the rudiments of the process is just developing.

As of the early 1990s, polled Americans showed concern for global climate change but ignorance of its nature, causes, and science's predictions of its consequences [Kempton 1991]. Lay causes commonly cited include aerosol spray cans and ozone depletion, general air pollution, and (correctly albeit exaggerated) deforestation [Kempton 1991, Meijnders 1998]. As consequences, interviewees in the early 1990s mentioned depletion of atmospheric oxygen, breathing of greenhouse gases, warm summers, sea-level rises, and especially the imposition of these problems on future generations [Kempton 1991].

The common public confusion of the problem with stratospheric ozone depletion, and the framing of solutions in terms of conventional end-of-the-pipe pollution controls, highlight the dominance of previously established conceptual categories ("mental models") in approaching new environmental problems [Kempton 1993]. Although respondents commonly showed a basic understanding of ecological species interdependence, they only weakly justified species preservation, and mostly for human use. They only vaguely appreciated the size of the potential impact of global warming on ecosystems and ecosystem services [Kempton 1991].

Most significantly, the public in the early 1990s did not understand energy efficiency's potential role in reducing carbon emissions; and even more fundamentally they failed to grasp the connection between fossil fuel combustion and climate change. (In open format questionnaires, CO^2 generation as a result of fossil fuel consumption was seldom mentioned as a cause of global warming [Meijnders 1998].) Reducing energy use evoked notions of decreasing energy services (cutback and sacrifice). Energy conservation in the US has a sullied image from the 1970s when it was associated with sacrifice and loss of amenities. In addition, the public conceptualized alternative energy sources more easily than energy efficiency. To increase the public's response knowledge, Kempton laid out the priority needs in public communications concerning climate change: Connecting climate change to energy use; concretizing understanding of energy efficiency; understanding that small changes in the mean global temperature could have huge effects; and realizing the sensitivity of the biosphere to climate change [Kempton 1991].

This apparent public ignorance is not overly surprising. Shove's invisibility premise holds that many environmental problems, including global warming, as well as most aspects of energy consumption, are invisible to the end-user, or at least have large "invisible" components [Shove 1991]. Knowledge is "made and mediated through modeling and measurement," but the end-user, and particularly the layperson, must make a "leap of faith" to accept energy consumption as a fact of reality. Connecting energy conservation actions to aggregate indicators of energy consumption requires yet another leap. The same holds true for connecting ordinary daily use and interaction with technology to individual energy consumption, to CO^2 emissions, and in turn to global warming and other environmental impacts (especially in the North).

Contrary data from Farhar [1994] suggests that awareness of the link between energy use and global warming had begun to increase by the late 1980s. This may help to explain a 1997 poll which found that a majority understood global warming very well (16%) or fairly well (45%) and still more recent polls showing an overwhelming acceptance of the (potential) reality of the problem, a relatively advanced understanding of the causes, and an acceptance of a certain level of costs for solutions (discussed shortly below) [Kull 2000]. To explain these results, one can speculate that either public understanding truly advanced over the decade of the 1990s, or the polls inflated answers by providing some of the missing basic knowledge as background prompts or through closed-format questionnaires.

Americans' *opinions and attitudes* towards climate change have also shown an evolution. In 1991, interviewees reacted very unfavorably to the idea of *adaptation* to the consequences of climate change without a *prevention* strategy, even when favorably framed. Many viewed adaptation as procrastinating and endorsing a do-nothing, business-as-usual approach. Kempton hypothesizes that economic discounting may not figure in the public's valuing an intact environment to pass on to descendants [Kempton 1991]. However, as will be seen, popular notions of what prevention or mitigation might entail seem too simplistic and largely technology-based.

In discussions of energy taxes, the public overwhelmingly believed their energy consumption (especially transport-related fuel demand) to be highly resistant to price changes [Kempton 1991]. A number of other studies have demonstrated that most of the public takes gasoline consumption to be highly inelastic and are therefore skeptical that increases in gasoline prices or taxes would change consumer behavior significantly[54] (cited in [Farhar 1994]). (The same (perception of the) non-elasticity in gasoline consumption has been found in the Swiss population (see [Dahinden 1997]). In 1994, though, the public reportedly preferred reducing energy demand to increasing supply, except if the energy were to be derived from renewable or alternative fuels [Farhar 1994].

As noted above, by the late 1990s polls showed a seemingly rapid development of understanding of certain aspects of the climate change issue. Accordingly, by 1999, the vast majority of the American public believed global warming was occurring or was in the offing and constituted a serious problem, even though a much smaller majority held that science had reached a consensus on the issue [Kull 2000]. Where opinion split was on the degree of the problem's urgency and therefore on the justification for taking immediate, potentially costly steps to address it. A minority favored taking costly steps, while the majority believed that the solution could be found in relatively low-cost measures – $25 per month per household in equivalent

[54] Following the huge spike in oil and gasoline prices in 2004, this elasticity will surely be re-tested through econometric analysis.

increased energy costs – and were unwilling to spend as much as $50 per month per household. Americans' technological optimism explains this reluctance and the finding that in fact a majority believed the necessary remedies would help, rather than damage, national economic performance in the short- to mid-term [Kull 2000].

If this survey is reliable, the majority of the American public subscribes very much to the mainstream environmental policy approach of limited technological fixes, or at least echoes this approach in their poll answers. The popular public conception that global warming can be addressed with a stricter application of the conventional regulatory end-of-the-pipe approach (and/or market-based mechanisms like emissions trading) is likely related to the public's grafting of the mental model of regional ambient pollutants onto this new problem. The popular opinion on the elasticity of gasoline consumption cited above shows the public doubts it is capable of significant changes in their levels of transportation demand (at least in the short to mid-term). Then the modest increases in fuel costs (e.g. in taxes) acceptable to the public could presumably only contribute to a "solution" to global climate change by enhancing funding for technology research and development, presumably for renewable fuels and enhanced efficiency. The studies cited in Chapter 2 cast doubt on the adequacy of such a solution, removed from any consideration of activity levels, to environmental problems characterized by both technology and consumption dynamics.

2.2.2 Psychological factors and knowledge in (energy-relevant) ecological behavior

2.2.2.1 Psychological factors in ecological behavior

A main feature of the Energy-Revealing approach to energy and the environment is providing people with the information and/or motivation deemed necessary (and often sufficient) to enable them to guide their behavior in an environmentally favorable direction. Information, knowledge, and motivation are examples of what some environmental psychologists term psychological variables or determinants of "ecological," i.e. environment-oriented, behavior. This section lays the ground for the investigation into the links between knowledge and behavior, and then between knowledge and social change, by giving an overview of relevant conceptions of ecological behavior and these psychological variables.

There is no model that provides a full explanation of the factors and influences on ecological behavior, energy-related or otherwise [Bolscho 1994]. Social psychologists link behavior to value orientations determined by beliefs, attitudes, and, circularly, behavior; developmental psychologists ascribe value orientations to cognitive and psychological structures; while educators link pedagogy to information and knowledge acquisition to concern and awareness [Finger 1994]. Many types of actions, not only ecological behavior, depend on knowledge and understanding in their different dimensions, ethics and values, attitudes, experiences, and emotions, all of which are tightly connected to one another through complex, multi-linked pathways [Aho 1998, Bolscho 1994]. Environment-oriented behavior, like energy consumption, exhibits no unified, non sector-specific master pattern and, as is widely recognized, there is no unequivocal, context-independent connection between environmental consciousness and ecological behavior [Bolscho 1994].

The theory of reasoned action or behavior (Figure 3-4) posits a dual chain of psychological influences on behavior intention, which in turn mediates ecological behavior. These psychological variables consist of factual environmental knowledge, attitudes towards ecological behavior, values, and norms (e.g. [Ajzen 1986] [Kaiser 1999a]):

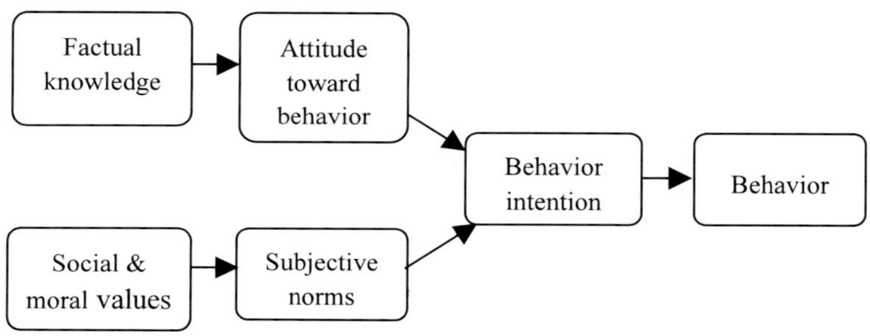

Figure 3-4. The theory of reasoned action

Environmental psychologist Kaiser's theory of planned behavior extends this by considering outside, uncontrollable influences as a random influence on specific behaviors [Kaiser 1999a].

Kaiser contends that a person's ecological behavior should be measured generally across the full spectrum of behaviors and that it should be assessed with reference to a group-defined measure of behavior difficulty. That is, behaviors can be ranked relatively according to difficulty of execution, based on a group's or society's frequency of performing them. By then adding influences beyond one's control, influences stemming both from general socio-cultural factors and randomly varying local or personal situational factors, Kaiser allows for inconsistencies in behavior: He adds the realism of permitting an environmentalist to fail to do the relatively easy task of recycling newspapers and the anti-environmentalist to abstain from car-driving, both for situational, and even personally transient, reasons. Provided situational factors are thus accounted for, and ecological behavior measured generally, Kaiser finds that ecological behavior intention, the last intervening variable, is empirically a strong predictor of ecological behavior [Kaiser 1999a].

Social dimensions add to the complexity of analyzing general and especially energy-related ecological behavior. Preventive responses to climate change are inherently collective, isolated individual behavioral change having little effect [Kempton 1991, Meijnders 1998]. Climate change is an eminent example of an externality and common property "tragedy," both socially defined concepts. Energy conservation behavior may follow the logic of game theory's "prisoner's dilemma" or psychology's "social trap," in addition to leading the individual into risk-benefit expectation traps and temporal and spatial discounting traps [Meijnders 1998].[55]

The social dimensions and general complexity ("messiness") of actors' energy use has made it difficult for psychologists to give generalizable, neat suggestions to policy makers. This has even prompted an exodus of psychologists from this research area [Shove 1998].

A different sociological approach is Gidden's social practices model.[56] Beliefs, norms, and values are part of the context of rules belonging to a shared social practice like heating, cooking, or traveling. The perpetuation of the practice depends on both these rules and on varying situational factors [Spaargaren 2000]. Giddens then defines lifestyle as the "set of social practices which an individual embraces, together with the story-telling that goes along with it" (Giddens cited in [Spaargaren 2003]). But social practices are also simultaneously influenced by institutional infrastructures, the collective socio-material systems of provision (of electricity, gas, and water, for example. See Figure 1.2). "When there is a high level – both in quantitative and qualitative respects – of 'green {i.e. environmental} provisioning' of a certain social practice, people are more or less brought into a position in which the greening of their corresponding lifestyle-segment becomes a truly feasible (and sustainable) option" [Spaargaren

[55] "However … greed is not the only motive underlying an individual's decisions in a social dilemma. Two more motives are important as well: the equity motive and the efficiency motive ... The efficiency motive implies that if individuals believe that the continued existence of a collective good is seriously threatened, they will aim to preserve it. Hence, if individuals believe that climate change seriously threatens conditions of life, they will be more likely to render support to mitigation policy" [Meijnders 1998].

[56] Spaargaren endorses Giddens' use of *social practices* and *lifestyle* as the units of analysis, instead of the individual, in research and policy for sustainable consumption.

2003]. This sort of treatment is more closely connected to the Social-Revealing and ecological modernization theories discussed below.

2.2.2.2 Knowledge

This section starts to explore the notion of environmental or energy "knowledge" in greater depth in psychology and other disciplines and contexts. The objective is to examine the variety, potential utility, and operation of knowledge in directing and mitigating the impact of resource and energy consumption.

The theory of reasoned behavior (above) suggests a complex process in which several variables intervene between knowledge and behavior. Thus, knowledge should be viewed as a "distal predictor of behavior that is conveyed by more behavior-proximal mediators" [Kaiser 1999b]. This means that several mediators closer to behavior convey, and attenuate, the influence of knowledge on behavior. As noted, these intervening variables may include incentives, intentions, attitudes, values, and other factors.

For the individual, environmental knowledge takes one of at least four forms: declarative, procedural, effectiveness, and social [Kaiser 1999b]. Most declarative environmental knowledge is factual apprehension, usually of the workings of environmental systems. Procedural knowledge has to do with effectiveness in the execution of a given ecological behavior. Understanding the relative effectiveness of alternative choices (e.g. of favorable consumption patterns in diet or transport) shows a level of effectiveness knowledge. Finally, social knowledge pertains either to pertinent social interaction skills or socially shared knowledge i.e. conventional or ethical social norms [Kaiser 1999b]. Environmental educators note that social normative knowledge and other types of knowledge develop together, suggesting that "moral and cognitive development go hand in hand" in environmental learning as well [Aho 1998].

More generally, a unit of knowledge must converge with other types of knowledge and often a wide array of other variables, both concurrent and distal, to exert a behavioral influence, and it is the proper convergence and not the quantity of knowledge that determines its effectiveness in promoting ecological behavior. Information or knowledge alone does not account for a very large portion of ecological or energy-related behavior [Kaiser 1999b],[Finger 1994]. For example, the social-cognitive process model of environmental behavior [Bolscho 1994] stresses the necessity of the convergence of "knowledges" in a conducive social context. It highlights the importance of social conditions "as a result of which the discrepancy between knowledge and action can be explained and perhaps overcome ...

Environmental behavior as a goal of environmental consciousness is dependent upon acquiring appropriate knowledge structures as well as being embedded in a suitable social and value context" (author's translation of [Bolscho 1994]).

Environmental knowledge acquisition appears to involve an intricate interplay of values, emotions, and other psychological variables. Meijnders' dissertation work investigated the influence of negative emotion on this knowledge acquisition [Meijnders 1998]. Working with Dutch subjects, she verified that strong environmental "threats" provoke negative emotions and tensions in the recipient, whose motivation and thought processes increase in an attempt to find a behavioral solution to reduce the tensions.[57] Emotional appeals concerning climate change may alter attitudes and opinions or evoke ecological behaviors, provided the mechanisms of anthropogenic climate change are clearly explained and assimilated; the contributory role of individual behavior is clarified; and arguments for mitigation measures are compellingly made. Otherwise, such appeals may only reinforce feelings of environmental "fatigue," helplessness, or fatalism [Meijnders 1998].

In some cases this fatalism may be deeply ingrained. In a Swiss study, Finger found that the strongest motivation for environmental learning is not necessarily acquisition of any conventional forms of knowledge but rather "psychological security" or "coping with fear." Many Swiss evidently learn about the environment mainly in order to know what they will have to face in the future. Thus, far from leading to heightened ecological behavior or social action for change, adult environmental education courses seem to substitute for such behaviors, by helping people cope with fears generated largely by environmental catastrophes as portrayed in the news media. Ironically, continued consultation of the media for more information leads to further fear, resignation, and substitution of learning for social action or activism [Finger 1994]. We conclude from this that if environmental fatalism is widespread or culturally endemic, it seems to act as a potent inhibiting mediator for (declarative) knowledge's conversion to proactive responses of any kind, except perhaps the collection of more of this knowledge. For climate change, this would implicitly endorse a politics of procrastination and lock in adaptation as the only response. Perhaps provision of a different knowledge set, or sort, could break this cycle. Socially-Revealed knowledge as one such potential sort is described shortly in the following section.

[57] The Social-Revealing and ecological modernization of consumption approaches would suggest that a person's possible behavioral responses are both direct and indirect, the latter spanning a range of consumer and political actions across different levels of integration.

The complexities of knowledge's connection to ecological behavior show the weaknesses of the conventional policy assumption of a straightforward link that can be exploited by public information campaigns and educational efforts. Although information has been one of the standard American energy conservation policy tools, providing information has not usually effected significant change in residential energy use: programs are beset with problems including credibility, failure to secure a behavioral commitment, or failure actively to involve the energy users. Informational programs often make the erroneous assumption, based on psychological or economic rational actor theories, that people will act on information about what steps they can take to save energy and money. Among other things, such programs fail to concentrate on the attractiveness, clarity, simplicity, or relevance of the message and/or the credibility of the sources [Stern 1987, 1992]. Nevertheless, information remains a popular policy tool in the public's eye. In focus groups with Swiss laypeople, asked to rank the relative acceptability of policy instruments – including information provision – as different means of achieving reductions in general energy use, subjects ranked information the highest for low and moderate reductions and equally high as other means for securing high reductions [Dahinden 1997].

Energy labeling of consumer appliances, seemingly a simple case of transmission of procedural or effectiveness knowledge, is likewise enmeshed in complex psychological and situational variables and communication issues. Energy labeling often assumes a strong link between knowledge (of the energy implications of consumer goods and services) and action (changes in consumer choice) [Shove 1997]. Yet, here too the amount of information may be less important than attracting the audience's attention and securing credibility. This depends on the presentation of the message, interaction between the sender and receiver, trust in sources, and many other factors discussed in the upcoming sections on communication.

Convergence with the proper social knowledge, and the proper social or cultural setting, may be important here, too. Cultural-sociological work in the area of the public's understanding of science suggests that ignorance of science may be a "function of active reflection upon, and construction of, the actor's social position and identity in relation to scientific-technical institutions" [Irwin 1996b]. Technological as well as scientific ignorance, even of the technical information a product label's designer intends to transmit, can be actively used for social purposes and may not always represent an intellectual vacuum (see [Michael 1996]). From a related approach, efficiency labels that implicitly encourage or even condone purchasing larger, or larger numbers of, electronic appliances, or using them more freely, perhaps in part because they are promoted as relatively energy efficient, may not serve the efficiency goals of the larger socio-technical system. Without a broader message (a constructive "shared understanding"), labels may validate symbolic or even counter-productive actions like purchasing relatively more efficient models and provide psychological salves instead of actually contributing to reducing overall environmental insults [Moezzi 1998].

2.3 The Social-Revealing approach and less discretionary influences on energy consumption

2.3.1 Theories, models, and advantages

The preceding sections surveyed several of the major items on the Energy-Revealing agenda, including public opinion and knowledge of energy consumption and the environment, and the meaning and effect of knowledge sets. This section takes up in greater detail the second, alternative approach to energy analysis and further develops the related concept of less or non-discretionary influences.

The Energy-Concealing/Social-Revealing approach to energy analysis, it will be recalled, takes embedded energy consumption in products, services, and systems for granted and instead tries to shed light on its socio-technical drivers and causes. A better identification and understanding of these practices helps to influence and manage energy-consuming services and devices. The hypothesis is if end-users are to have a role in managing and shaping their own energy-consuming practices, services, and devices, they require a broader accounting (i.e. knowledge and understanding) of the determining factors than provided by the individual-oriented Energy-Revealing approach alone.[58]

Chapter 2 cataloged the role of social, economic, institutional, and other external factors on environmentally significant consumption. The following discussion focuses on their treatment for energy issues by several different theorists: Kaiser's situational variables are related to the notion of non-discretionary energy-determining factors. Stern's model of the causes of environmental and energy-relevant behavior makes something of a transition to (Energy-Revealing)/Social-Revealing: It keeps energy use visible but starts to illuminate social and institutional factors that are largely beyond the end-user's direct control. Shove, Lutzenhiser, and Wilhite's sociological perspective embodies the Social-Revealing approach itself. The final two themes, the public's influence on energy-relevant decisions and perceptions

[58] The software developed for the interviews described in Chapter 4 shows the comprehensiveness and synergies possible when the two approaches are combined.

of control, hint at the potential and complications of the use and communication of this Social-Revealing approach with laypeople, issues taken up again in the concluding section of this chapter and the succeeding chapters.

The reader will recall that one type of Kaiser's situational factors consists of non-volitional, social and cultural constraints on ecological behavior that partly account for the differences in the ease with which various behaviors can be carried out [Kaiser 1999a].

Kaiser divides situational factors into two types: (1) universal, public constraints measured by a statistically determined, comparative level of difficulty of different behaviors, and (2) (stochastic) personal constraints on any individual's difficulty in carrying out behaviors, stemming from randomly varying life situations [Kaiser 1998b]. The first factor type imposes general constraints, enabling or making difficult a given behavior for everyone in a regionally or culturally defined group. However, in the face of greater or more severe situational constraints on ecological behavior, the more one overcomes these constraints the more one can be said to be behaving ecologically, and the more one singles oneself out as an outlier in a distribution curve of behavior frequencies across difficulties [Kaiser 1999a, b].

In one sense, Kaiser's situational factors elaborate the concept of a discretion continuum by putting the universal constraints on a scale determined by the distribution of difficulty in performance across people in a society. Furthermore, he terms the personal factors situational (non-discretionary in a sense), and models them as random, thereby allowing for inconsistency in action. On the other hand, however, as a psychological construct, this conception of situational factors cannot easily accommodate certain aspects of the socio-technical, Social-Revealing perspective presented below.

Stern and Oskamp's model of the causes of environmentally relevant behavior (Table 3-1) is often cited as a particularly good illustration of structural and institutional constraints on (energy-oriented) ecological behavior [Stern 1987, 1992]:

From a version of this model, Stern has described a pyramid typology of energy consumption influences according to the degree of direct control by the end-user [Stern 1992]. At the lowest level are *direct energy-using activities* like turning on a light or using an electrical appliance. The middle level consists of residential or industrial energy users' *technology choices*, whose effects on energy consumption are more indirect, e.g. the choice to settle in a house of a given size and location largely determines one's heating and commuting needs. The intensity of use and the longevity of the

Table 3-1. Stern's Causal Model of Resource Use*

LEVEL OF CAUSALITY	TYPE OF VARIABLE	EXAMPLES
8	Background factors	Income, education, number of household members, local temperature conditions
7	Institutional factors	Owner/renter status, direct or indirect payment for energy
6	Recent events	Difficulty paying energy bills, experience with shortages, fuel price increases
5	General attitudes	Concern about national energy situation
	General beliefs	Belief households can help with national energy problem
4	Specific attitudes	Sense of personal obligation to use energy efficiently
	Specific beliefs	Belief that using less heat threatens family health
	Specific knowledge	Knowledge that water heater is a major energy user
3	Behavioral commitment	Commitment to cut household energy use by 15%
	Behavioral intention	Intention to install a solar heating system
2	Resource-using behavior	Length of time air conditioner is kept on; Insulating attic, lowering winter thermostat setting
1	Resource use	Kilowatt-hours per month
0	Observable effects	Lower energy costs, elimination of drafts, family quarrels over thermostat

*From "Managing Scarce Environmental Resources" (p. 1063) by P.C. Stern and S. Oskamp, in D. Stokols and I. Altman, *Handbook of Environmental Psychology*, New York: John Wiley & Sons, Inc.

infrastructure influence the long-term impact or contribution of these choices. Stern calls the highest, most indirect level of influences *policy choices*: "Decisions by manufacturers of automobiles, appliances, or industrial equipment about whether to produce highly energy-efficient products are energy policies in that they constrain consumers' technology choices" [Stern 1992].[59] Tax deductions for mortgage interest payments and subsidies for highway construction are similar government policies. They provide consumers and companies with overarching incentives,

[59] Similarly, "Major technological innovations are roughly equivalent to governmental legislation, in the sense of establishing an enduring framework for everyday life" [Swearengen 2003].

disincentives, opportunities, or non-opportunities. Although such policies are set by corporate boards of directors, legislators, and politicians, they are sometimes amenable to outside input or public opinion.

Stern emphasizes that influencing business and government policy makers is more effective than individual behavioral change at addressing energy conservation and many other environmental problems. Technology choices and policies should be targets of energy conservation efforts over and above everyday energy-relevant behavior, according to the prevention principle of pollution and energy use. This approach is supported by the longevity and impact of technologies and the tendency towards short-term fluctuations (reversals and re-reversals) in energy conservation behavior [Stern 1992, 1997].

Stern's energy choice typology is clearly on the conceptual way towards the Social-Revealing perspective on energy consumption. Both emphasize the determining decisions of institutional actors like building contractors and developers, manufacturers and retailers that frame or limit choice in end-users' decisions. Yet the full-fledged sociological-technological approach also incorporates analysis of the evolution of social norms with technologies, choice thresholds and technological "lock-in." "... Individual choice in industrial societies is limited by the way cities, energy and water supply systems, housing designs, product designs, etc. are configured. Individuals can influence what happens at the end of the pipe, but significant changes in energy use are bounded by the 'upstream' systems they are plugged into" [Wilhite 2000]. The evolution of such systems, the roles and interaction of agents and actors involved in their development, and the possibilities for modifying or directing the course of development are all important for studying and reducing energy demand.

Reflection on the history of technology, policy, and social choices can increase social "self-awareness," for example in the area of housing and office buildings: "The built environment embodies the expectations of designers, financiers, and occupiers, while also reflecting the organization and structure of the construction industries. As such it fossilizes past patterns of social relations and creates a form of inertia that slows changes in energy and resource use" [Shove 1998]. Historical examination can help identify important decisions, moments, and thresholds for technological and societal change as they arise, improving the quality of the long-term social investments they represent and possibly avoiding lock-ins to unfavorable energy outcomes [Wilhite 2000].

Historical analysis of this type may improve future group decision making and proactively warn against further unfavorable lock-ins, but can past technology choices be revisited, and past choices retaken or reversed? Bijker's answer is generally, but not absolutely, no. In technological development, "closure" of competitions between alternative technological artifacts establishes the victorious and henceforth dominant form of technology. Before closure, technological artifacts exhibit "interpretive flexibility," each relevant social group interpreting and attributing meaning to (constituting, even) the artifact. After closure and especially after degrees of "stabilization," interpretive flexibility is greatly limited or altogether impossible, and in fact with time seems retrospectively not to have been present because the history of the development is rewritten. Reopening the field to socio-technical alternatives – reversing closure – is theoretically possible but practically quite difficult. Theoretically, the heterogeneity of the factors involved – social, psychological, technical, and normative – leaves open the possibility of reversing the closure process by allowing for more or less advanced states of technological stabilization. But in political terms, highly stabilized technologies exhibit "obduracy" due to successive attributions of meaning, building up of structures, and hardening of supportive networks of practices and social institutions. Closure and stabilization represent increasing levels of semiotic power,[60] as more groups and players are drawn in to the artifact's "technological frame" and have stakes in defending and elaborating on the artifact's settled meaning [Bijker 1995].[61]

Returning to the explication of the full-fledged Social-Revealing approach, analysis of the interlocking development of technologies and social norms is also necessary. Technologies (and people involved in diffusing and marketing them) sometimes generate new social "needs" and may conveniently pose themselves as the answer to these needs. An example is standardized American air conditioning uniformly installed in houses regardless of actual climatic need and tuned to provide a certain, perhaps unnecessarily low, indoor air temperature. This may create technology

[60] Semiotic power: The order of the normal, the habitus, "fixed and represented in technological frames" [Bijker 1995].

[61] This digression into social constructivist theory suggests the following series of analogues for the discretionary/non-discretionary dialectic from political and social theory. They are especially appropriate when applied to examinations of the historical evolution of technological frames or challenges to the prevailing technological regime in a political context:

- discretionary vs. non-discretionary
- voluntaristic vs. structural
- change vs. constancy
- malleability vs. obduracy
- micropolitics of power vs. semiotic power (Bijker)
- transformative capacity vs. domination (Giddens)

supply-dependent norms of temperature comfort and fuel demand where and when previously they may not have existed – the so-called "manufacturing of demand," often followed by the seemingly immovable entrenchment of this type of demand in society. Other readily available, more environmentally friendly means of providing services like climate control (e.g. daylighting techniques) "may not be socially viable, given the layers of expectation and practice now built up by and around this particular technological option" [Shove 1998]. For instance, time-saving and convenience norms, in private life as well as business, drive increasing substitution of machines for human labor and the demand for farther and faster travel, both of which have profound impacts on energy use [Wilhite 2000].

Compared to psychology theory, the Energy-Concealing/Social-Revealing approach more readily allows for social renewal and ecological modernization to be incorporated. Social-Revealing better accommodates the idea of different option spaces, for example advanced technology, alternative technologies, and alternative social norms,[62] all of which may have previously existed (in terms of social norms), may exist in other societies (apparent through cross-country comparisons), or may not yet exist (energy efficiency improvements beyond the current technical potential or different structural economic arrangements (e.g. promoting restraint and sufficiency). This approach accommodates larger changes in higher-level system barriers, social norms, or "absolute" technological limitations that can be overcome by suitable long-term private and public investment.

Insights from the Social-Revealing approach, being sociologically rooted, could be misinterpreted as an excuse for denying most elements of choice and therefore responsibility in individual energy consumption. In bringing such an institutional perspective to the end-user, this is not at all the intention. On the contrary, showing what is legitimately outside one's direct control (non-discretionary) highlights by contrast what *is* under one's control (discretionary), although Kaiser shows the distinction depends strongly on personal characteristics vis-à-vis society as well as random influences that require a specific situational context for any coherent analysis. Balancing this approach with elements from the conventional Energy-Revealing economics-engineering approach – those that concentrate on direct contributions to energy consumption and associated environmental impacts – helps some people to take personal responsibility where appropriate while peaking their interest in less discretionary social and technical issues as well.

[62] See Chapter 2, Section 3.3.2, "Anthropological/sociological treatments."

However, a separate argument runs that leaving energy embedded in social functions and avoiding the explicit focus on energy and the environment characterizing Energy-Revealing can bypass the linearity conventionally assumed in the environmental knowledge → understanding → action policy process and come directly to the action stage, while at the same time possibly enlarging the pool of stakeholders for this stage. Moving stakeholders or audiences up the "knowledge infrastructure" (see [Connors 1998]) directly and sequentially, and for climate change alone, to the consumption-related elements of a policy "solution" is a laborious process, as the slow public diffusion of knowledge about energy efficiency and climate change has demonstrated. Instead, as documented in Chapter 2, there is a collection of concerns which may have overlapping solutions in the form of energy conservation and source reduction: e.g. climate change, national security, national competitiveness, reduction in materialism, psychological or social renewal, and so on. For policy applications, approaching energy and environmental problems from the non-discretionary and Social-Revealing vantage point may bring different audiences to the forum and allow multiple constituencies for various (but related) problems to rally around common measures for their amelioration.

2.3.2 Public perception and perceived control of less discretionary factors

This section offers a preliminary look at the degree of the public's perception of less discretionary factors and its perceived degree of indirect influence on them.

Farhar notes that American energy consumption policy shifted in the 1990s from a focus on behavioral changes to limited institutional changes such as demand-side management, appliance standards, and energy-efficient mortgage lending. This shift appears to have had the full support of the populace [Farhar 1994]. In polls, Americans mentioned high institutional barriers as reasons they do not drive less, carpool more and use public transportation more frequently: 34% responded "I'm doing all I can," 40% objected to high up-front costs of energy efficiency improvements, and 16% cited inconvenience or difficulty in changing habits [Farhar 1994]. Farhar goes on:

> It seems reasonable to speculate that the public wants efficiency and the use of renewable energy to become "institutionalized" – to be built into the routine way of producing cars, buildings, and energy services, for example. The public seems willing to change its behavior, up to a point. Then much of the public seems to want institutional change, so

> that efficiency and reliance on sustainable energy sources are normal, not special activities … The public wants the burden of change to fall on institutional (not just on family and individual) shoulders. Institutions include governments, utilities, automakers, builders, the financial community, and others whose policies, procedures, and practices constrain individual and household energy choices. Consumers say they will pay through mechanisms that support institutional change – such as "green pricing," "gas guzzler" taxes … and DSM programs – if the results are truly beneficial to the environment [Farhar 1994].

One notices the departure from the assumption of the Social-Revealing approach that significant changes in aggregate energy use and general environmental protection "will be predicated on a significant social transformation." These comments suggest relatively modest institutional change. For example, the populace portrayed here seems unaware of the importance of feedback from technological and institutional change on group lifestyles, norms, and normative individual practices. "Institutional change" in this usage seems rather shallow; for example, it does not address the interlocking development of technology and energy-intensive lifestyles (see [Wilhite 2000]).

It is interesting to compare Farhar's account of Americans' perception of the role and nature of non-discretionary factors to Finger's concurrent analysis of the Swiss public's perception of these factors:

> When asked what ... was preventing the solution of today's environmental problems, only 9% indicate a lack of information and knowledge. 44% ... say that economic stakes were too high, 20% state that political stakes were too important, and another 15% indicate too much individualism ... In short, the Swiss have a quite disempowering perception of environmental issues and problems, which they mainly see as global, urgent, and somewhat overwhelming. They are afraid of them and pessimistic about the future. They see major economic, political, and cultural forces in society preventing environmental problems from being addressed [Finger 1994].

It appears from this that many Swiss believe they understand these non-discretionary factors all too well but lack a conviction in their own, or their government's, ability to change them; they feel politically, and geo-politically, helpless. If Farhar's analysis is correct, it seems that in comparison Americans, despite common rhetoric deprecating "big government," leave the main work in confronting energy-related environmental problems to their government or governmental guidance, and seem to trust in its ability to solve them. Not so many of the Swiss. Finger suggests that they are fatalistic, emotionally paralyzed, or so pessimistic and disempowered by their awareness of the very non-discretionary nature of factors driving environmental problems that they content themselves with information gathering as solace against future ecological hardships.[63] In this case, further information related to energy consumption (in the context of environmental degradation) does not seem likely to provide the missing incentives or motivations; this situation calls for political rejuvenation on a grand scale. However, the problem of disaffected and disempowered electorates in Western democracies is largely beyond the scope of this book. (We return to the question of public perceptions and "use" of scientific and social knowledge at the end of section 3.2.4 on risk communication below.)

3. ENERGY AND RISK COMMUNICATION

3.1 Public communication

Having explored various types of energy knowledge and touched on some of the necessary conditions or co-factors for their use, we turn fully to the issue of their (bi-directional) conveyance through communication. Just as the standard energy consumption messages can benefit from a broadened perspective, so can the conventional means of communication. The later discussion on risk communication extends the discussion to a wider range of actors and broadens the notions of communication flows, channels, and participation.

Developed especially in connection with the nuclear energy industry, public communication (PC) of science and technology has more recently come back into vogue as a tool for affecting public attitudes on complex scientific issues. It is regarded as most effective when used in tandem with other tools like financial or regulatory measures [Meijnders 1998]. Its effectiveness also hinges on the concurrence of features examined below.

Covello's popularly cited definition of public communication in this context, sometimes used to define risk communication as well, is "any intentional exchange of scientific information among affected parties on health or environmental risks" (cited in [Kant 1995]). With regard to energy this reads "any intentional exchange of scientific information among affected parties on energy policy dilemmas." This usage does not include all aspects of communication with lay audiences, even of strictly Energy-

[63] Some of the interviewees in Chapter 4 aired views that challenge these generalizations.

Revealing issues. Typically contrived functions for public communication include informing, educating, managing and resolving conflicts, and facilitating decision making. There is a clear demarcation between PC for general knowledge enhancement, expected to affect beliefs, and PC intended to change a specific attitude or behavior [Kant 1995]. (The remaining discussion in this section draws on [Kant 1995] unless otherwise cited.)

Previous sections have described the multi-variable complexity of inducing behavioral change. In the context of communication, necessary co-factors include audience attention, comprehension of the message, suitable emotional appeal, and alignment of attitudes, but even then the durability of any induced change is questionable.

In the Netherlands and other Western European countries, public communication in the service of decision making on issues of interest or import to society is supposed to "broaden the societal consensus for political decision making." Yet the broad participatory mode of communication has not necessarily been the rule, even in Western democracies.

When the classic communication models were developed after WWII, authorities were seen as initiating uni-flow scientific communications in order to increase levels of knowledge, assuming that this would lead to desired actions on the part of the receivers. In the stereotypical public information campaign, experts on rational energy use approach the matter from a technical standpoint and address an ignorant and emotional public. The increased knowledge imparted is supposed to palliate the public's negative emotions.

According to the traditional engineering model of information transfer through public communication, developed in the late 1940s by Shannon, Weaver, and, separately, Lasswell, communication consists of source, objective, message, medium, transmission, channel, reconstruction-decoding, audience, noise, and other elements [Kant 1995], [Renn 1991]. The original model is static and mostly unidirectional compared with current models. Somewhat later models of public communication as diffusion of information to produce behavioral changes added attention, understanding, attitude, intention, behavior, and maintenance of the behavioral change. Attention and attitude, which depend on one's personal situation, are recognized as limiting factors for PC efforts.

Depending on the actors' implicit assumptions and the operational rules, the relationship between senders and receivers in public communication of science in the West has tended to be dominated by one of four approaches: Technical, market, justice, and participatory. The *technical* approach involves technicians primarily and an emotional public only secondarily. Provision of public information is separated from decision making. The *market* approach deals with implementation of policy or technology "as a process of negotiation in a market situation." The *justice* approach emphasizes an equitable distribution of risks determined by pre-established procedures and bargained positions. In the participatory approach, public communication centrally involves information and education of the public. A well-informed, participating public is considered necessary for proper decision making and to avoid a reduction in the vigor of the democratic

---

The classic communications sender-receiver model

The following reviews some basics of the updated but conventional communications engineering (sender-receiver) model (based on [Kant 1995]. The risk communications concepts presented in the next section enrich or reformulate aspects of this model.

According to the general sender-receiver model, characteristics of the message include "correctness," completeness, comprehensibility, belief, trust, attention-getting, and ethical dimensions. Information needs to be relevant and relatable to local circumstances, otherwise it will not be credible or effective. Objectivity may be enhanced by increasing the number of sources and including and explaining contradictory views [Schneider 1993]. Written media may be preferable in cases of complex information or with sources whose credibility or attractiveness is dubious.

Although variously posed as the neutral scientific or altruistic communicator, the objective, disinterested sender (source) does not exist. The source is expected to have an interest in the communication outcome, and a lack of stated goals arouses suspicions, at least of incompetence. It may be preferable for the source to reveal objectives explicitly rather than to have the audience make poor assumptions, since they will draw conclusions in any case.[64] The source's credibility depends partly on the quality of expertise and is particularly important when technology is involved. Attractiveness depends on the convergence of attitudes and sympathy and familiarity with the source. (Trust and credibility will be expanded upon in the upcoming section on risk communication.)

The receiver is also known as the audience or target. It is commonplace to hear that the contents and processes of the communication must be matched to fit the target group (e.g. [Aho 1998]), although this is almost self-evident. The receiver's characteristics include knowledge level, involvement (engagement) with the issue, expectations, needs, and motives. Parameters of the audience's "needs and wishes" include involvement, educational level, extent of (cognitive or emotional) removal from the risks involved, and vulnerability.

Education is an important influence on the audience. The complexity of environmental and energy issues increases the importance of the difference in levels of education among members of an audience. Finally, as with knowledge acquisition in general, social issues surrounding the receiver – his social position, support, networks, and the like – are often important for processing, digesting, and potentially acting on the message of the communication.

---

[64] This was the route taken in explicitly specifying the response framework for the interview participants (Chapter 4).

process, much of which stems from the public's exclusion from and ignorance of scientific and technological policy matters. Dialogue is the dominant interactive mode. This is an idealized typology, and the reality is not so easily differentiable, as most public communication involves elements from all of these models.

3.2 Risk communication

In the risk analysis paradigm used by environmental and public health regulatory agencies, internal risk communication (RC) permeates the process of communication between professional risk managers and risk assessors. External risk communication, on the other hand, is meant to obtain information and feedback from stakeholders; inform and educate them; enhance trust and credibility; change behavior or prompt protective actions; and/or warn about disasters and emergencies (Yoe 2003). In its broader usage, risk communication can be considered a subset of general public communication and can also be defined as an exchange of information about risks related to the environment or human health. The risk communicator is defined as "the individual or institution that intervenes to change existing knowledge or perception" [Kasperson 1991]. Traditional objectives of RC include producing changes in knowledge or attitudes, as in PC, prompting individual or group protective measures, or fulfilling other functions related to risk management, conflict resolution, or public participation. Risk communication has recently broadened its scope to include analysis of the exchange of information of certain kinds among multiple social subsystems (see [Renn 1991a]). Several relevant aspects of these types of exchange are taken up below.

3.2.1 Criticisms of conventional notions of risk communication

The communications engineering model of receiver and sender was once a dominant model in RC just as it was in general PC. This model was criticized as a social engineering version of a marketing approach whereby senders intentionally provide persuasive information in a manner targeted to receivers and measure the success of their efforts by the degree of behavioral change it prompts. This approach was inadequate for the analysis or design of risk communication. For one thing, it tended to be removed from the socio-political context and varying perceptions of the problem: The institutional setting for the communication is an important part of the experience and result. Risk communication is part of a complex communications web in which various groups and cultures possess varying perceptions, values, and interpretations [Kasperson 1991].

The other once prevalent model now often criticized is that of the altruistic communicator bent on advancing the public good. Political aspects of the risk communicator are downplayed in this model. The altruistic communicator is traditionally seen as "striving to manage risk in the 'public interest' and beset by a host of problems relating to the complexity of risk information, uncertainty, opaque social values, unscientific media, and a disinterested or volatile public, {who} must somehow produce in the 'recipients' a grasp of the scientific facts and an ability to put risks in perspective" [Kasperson 1991]. This model was taken to task for focusing on problems and failings in the target groups while neglecting problems with the risk communicators themselves. Risk communicators reflect the institutional goals and professional cultures of their job settings. Unstated motivations underlie the communication and affect and possibly confuse the audience, as noted in the previous section on public communication.

More modern theories of RC, in improving on some of these defects, nevertheless recognize the complexity of the processes at work and the limitations of their own models. RC literature cannot provide hard and fast guidelines for communicating with the public. Empirical evidence as a whole shows that "individuals as well as social units make use of a complex variety of internal and external cues to process messages and that the variation of one or two factors may only lead to marginal changes in the outcome" [Renn 1991b].

3.2.2 Levels of trust in management of "creeping danger" risks

Anthropogenic climate change, as a modern systemic risk, exhibits complexity, uncertainty, ambiguity, and open system boundaries. Uncertainty stems from deficiencies in (scientific) knowledge. Complexity denotes the presence of numerous intervening variables between cause and effect whose connections often need extensive modeling to describe. Laypeople are typically thought to be averse to exploring these connections. Ambiguity stems from heterogeneous values among the various stakeholders. With ambiguous risks, even ostensible benefits, not just potential hazards, may be contested (e.g. GMOs or high levels of consumption). Open system boundaries refer to the amplification of risks originating in the purely (bio)physical realm into the political, economic, social, and psychological realms [Renn 2004]. BSE ("mad cow disease") and its human variety are currently being highly socially amplified, whereas at least in the past, perhaps because of its high unfamiliarity, the risk of climate change seems to have been significantly socially attenuated. As it has shifted onto the international agenda, the risk of climate change has been amplified by its imperceptibility, increasingly validated human ("artificial") provenance, its involuntary and seemingly uncontrollable nature, the lack of clear risk managers in which to place one's trust, and its inequitable global and intertemporal distribution of costs and benefits.

In terms of the ways that people perceive and intuitively evaluate risks, climate change, like many food borne hazards, falls into the category Renn calls a "creeping danger" (others have called it a slow bomb). Creeping danger risks exhibit high complexity, delayed effects, and a high degree of invisibility: the dangers are only clearly apprehensible through expert information and modeling. When people are faced with an invisible hazard, which leaves them totally reliant on information from third parties, their willingness to accept the risk depends greatly on the trust and confidence they place in its risk managers (commodities in short supply in today's Risk Society). In the absence of any trust, people have no tolerance (they demand "zero risk"). If they are undecided as to whether they can trust the risk managers and the information provided to them, they rely on peripheral signals from the risk communicator and other external criteria (e.g. his source of funding or ideological orientation) [Renn 2004].

3.2.3 Routes of persuasion and levels of debate

Thus, Renn (1991b) envisions two routes of persuasion in risk communication, basing them on the psychological elaboration-likelihood model. A receiver of RC uses the central route when he treats the substance or content of arguments and considers the pros and cons of issues in determining whether or not to believe in a message from a risk communicator. He uses the peripheral route when non-substantive cues, for example the risk communicator's prestige, help him assess the credibility of a message or its sender. A peripherally engaged recipient places trust "holistically" in the sender, whereas the centrally engaged receiver decides whether or not to invest trust for each specific message from a given sender. Most issues in modern, specialized societies are too complicated for good analysis based on personal experience or knowledge and the plausibility of arguments; thus, people tend to use the peripheral route more often. Renn notes, however, that a delivery style geared towards generating interest in a message through the peripheral route by use of cues can be offensive to individuals who take a central interest in the matter.

Most information programs of the environmental or Energy-Revealing mold implicitly assume a mostly central RC processing route and a relatively high level of public trust. If they are couched in environmental terms or put into the climate change context, the complexity and controversy involved may, however, make lay receivers unsure as to the reliability of the message and make the peripheral route more important. Or, with the continued decline in the public's trust in institutions, and, for some, science and technology in general, peripheral RC may too become less effective in engaging people on environmental issues as they lose their tolerance for the risks involved.[65] Approaching energy consumption from the starting point of wider contexts or other issues, as is possible through Social-Revealing, and using individuals and small groups as risk communicators rather than institutional spokespeople, may help re-engage audiences through one or the other route.

Risk communication can be categorized according to levels of debate and discourse as well as routes of communication, depending on the nature of the risk: (i) complex but routine risks, where science can resolve the uncertainties; (ii) uncertain but non-controversial risks; and (iii) highly contested (ambiguous) risks [Renn 2004]. Each of these levels has different requirements for involving stakeholders in a discourse about the risk. The first level (i) requires an epistemological discourse, a factual dialogue primarily among scientists concerning probabilities and potential damage from the risk in question (e.g. health effects of air pollution). The second level (ii) needs a reflective discourse among scientists and other stakeholders, a "clinical mode" that deals with risk management, responses, and experiences of institutions (e.g. regulation of salmonella bacteria levels in raw and processed meat). The third level (iii) demands a larger discourse, even a "world view perspective," on values, lifestyles, or fundamental trade-offs (e.g. technology's place in society, environmental justice). As one moves up these levels, the degrees of complexity and intensity of conflict increase. Renn concludes that the type of concern of the audience must determine the level at which the RC message operates and its corresponding content [Renn 1991b, 2004].

Several factors, including declining public trust in institutions and products of new technologies like genetically modified organisms and nanotechnology,[66] could push risk debates from the first to the second or third levels. If risk debates are held on the third level, trust is not furthered by technical arguments or institutional competence. Here, trust depends on a consensus on basic issues often having to do with lifestyles and values, a consensus that is clearly lacking in pluralistic Western societies. This third level involves a "macro-sociological framework" that resists empirical testing and generalization across cases [Renn 1991b]. For example, it seems that conventional energy analysis (Energy-Revealing) generally presumes a

[65] "What scientists interpret as a naïve and impracticable public expectation of a zero-risk environment can thus be seen instead as an expression of zero trust in institutions which claim to be able to manage large-scale risks throughout society" [Irwin 1996b].

[66] More generally, public understanding and acceptance of converging new "NBIC" (Nano-, Bio-, Info-, Cognitive) technologies)

type of risk debate and communication centered around the first two levels and rejects attempts to take the discourse explicitly to the third level; whereas Social-Revealing tends, or lends itself, to third-level debate. The third level also lacks a clear communication medium [Renn 1991b]. The development of the enhanced household energy software tool (Chapter 4) was meant to contribute to the furthering of such a medium for risk communication on issues of energy consumption and climate change.

3.2.4 More on the social amplification of risk

The social amplification of risk depends on multi-level exchange and feedback. Individuals' perception of and reaction to RC from people and institutional senders in turn spur institutional responses or changes [Renn 1991a]. That is, secondary effects of the risk communication may involve "changes in social and institutional behavior that were induced by individual responses to risk communication" [Kasperson 1986]. "Secondary effects evolve as responses to signals sent from the receivers of risk communication ... to risk handling institutions. Secondary impacts include such effects as enduring mental perceptions, personal apathy, political pressure, institutional or political changes, and new social movements" ([Renn 1991a], emphasis added). Secondary impacts are communicated to individuals and social groups, and the resulting new feedback may generate tertiary social amplification effects of the original communication. On a macro level, amplified RC interacts with many other influences on general social communication and decision making [Renn 1991a].

To charge the social amplification process, media and other institutions must eventually take up the topic of concern of individuals and especially groups. Social groups and institutions are major agenda setters, even when the original impetus, and the distal source of institution-changing secondary effects, comes from individuals [Renn 1991a].

The social amplification of risk framework is inclusive in that it permits analysis of communication on an individual (micro), group (meso), and social (macro) level, and all combinations of the three. In the matrix presented in Table 3-2, each cell embodies a communication situation wherein sender and receiver may be an individual, group, or political institution.[67]

[67] In row 3, one of the motivations for RC from government agencies is to build public confidence in their assurance that the political and regulatory system can manage environmental risks, showing evidence of the openness of the decision making process, its flexibility, and resilience [Renn 1991a]. However, in today's Risk Society this function is increasingly unfulfilled. It is partly the failure of governments with respect to climate

3.2.5 Proposed application of risk communication and ecological modernization, with caveats

Renn's social amplification of risk is one of several frameworks that lend themselves readily to describing possible research and policy applications of this book. The empirical study, based largely on the research of a few individuals at the Swiss Federal Institute of Technology, can be made part of risk communication to individuals from a scientific institution (Table 3-2, cell (row 2, column 1)) and private individuals (1,1) ultimately to encourage secondary communication responses to groups and institutions on their parts, in their capacities as individuals (1,2) and (1,3), and maybe even as groups (2,2) and (2,3). Specifically, communication about less-discretionary energy use can inform (and motivate) people to participate in a dialog that promotes secondary RC effects constructive in ameliorating environmental damage and climate change. Primary RC concerning discretionary energy use, on the other hand, focuses on the communication pathways captured in cells (2,1) and partly (1,1), mostly for the purpose of informing and motivating individuals to modify their own risky direct energy consumption behavior. In Renn's terminology, the ultimate policy objective for non-discretionary communication would be to support the generation of secondary institutional effects of RC to individuals, positive effects on key organizations and institutional arrangements that lessen or mitigate the risks they pose in their current form to the global environment.

Knowledge's potential to facilitate the process of amplified risk communication and secondary effects, especially from individuals acting as citizens, consumers, and social actors to support environmentally favorable institutional and cultural changes, is simply risk communication terminology for what in ecological modernization theory is called monitoring for consumer-citizen involvement in the ecological modernization of consumption.

The most important but unusual outcome of educating consumers about energy consumption is their greater input into government and corporate (technological) policy choices normally largely closed to them. This assumes the appropriate political and corporate communications channels are in place. The explosive development of the Internet seems promising in this

---

change prevention/mitigation and the general invalidity of this assurance for environment risks that suggests the approach of using (grassroots) individual leverage on governments to promote institutional changes.

Table 3-2. Renn's [1991a] objectives of risk communication, levels of analysis, and pathways for the social amplification of risk

From/to	MICRO-LEVEL	MESO-LEVEL	MACRO-LEVEL
	Individual	**Group**	**Society**
Individual	Persuasion for risk reduction	Influence on group decision	Change of risk policies
	Risk acceptance	Request for support	Request for support
	Education	Education	Information
Group	Education	Education	Information
	Support	Coalition	Influence on risk policies
	Persuasion for risk reduction	Conflict resolution	Acquisition of social resources
	Risk acceptance	Prestige	Change in risk culture
	Acceptance of risk management	Acceptance of risk management	Compliance with risk standards
	Trust in group's competence	Trust in group's competence	Development of incentives for structural change
Society	Education	Education	Strategies for risk management and regulation
	Risk reduction	Risk reduction	Agenda for risk agencies
	Emergency response	Emergency response	Institutional reform
	Acceptance of risk management	Acceptance of risk management	Development of new paradigms of risk
	Trust in risk agencies	Legitimization of risk agencies	Changes in risk culture
	Loyalty with respect to the risk-handling capacity of society	Loyalty with respect to the risk-handling capacity of society	Influence on international and global risk policies
		Mediating in conflict resolution	International conflict resolution

regard, for the greatly increased information and transparency it provides users regarding the details of these choices in industry and government and the general ease of information and opinion exchange it facilitates.

This outcome fits well with the "politics of technology" described by the theories of Science-Technology-Society and the Social Construction of Technology (SCOT), which "stress the malleability of technology, the possibility for choice, the basic insight that things could have been otherwise." Demonstrating the interpretive flexibility of technological artifacts highlights the political nature of past technological choices and debunks the deterministic spin retrospectively cast on the history of their development by political and business interests who benefit the most from the status quo (the "winners"). Thus it offers a basis for wider democratic participation as on-going or new socio-technical choices present themselves [Bijker 1995].[68]

We note in passing that enhanced lay knowledge of the environmental-, energy-, or society-related dimensions of energy consumption also seems to conform well with the participatory mode of communications and its traditional aim of improving the basis for decision making in democratic processes. Lay input into risk communication and management also seems all the more warranted in an age of declining trust in, and even alienation from, major state and corporate institutions.

As noted, this application can also be described in terms of ecological modernization (EM) theory, especially recent Dutch sociological work. As a social theory, ecological modernization concerns itself with the "(re)design of central institutions of modernity in dealing with the ecological crisis and on the basis of environmental criteria" [Mol 1995]. The reflexive variant (Hajer in [Mol 2000]) has broadened its focus to the "structural change of socio-technological systems" [Mol 2000].[69] This implies that the state provide favorable conditions and contexts for consumers (and producers) to take a larger role in environmental improvements [Mol 1995].

The theory of the ecological modernization of consumption looks favorably on what seem to be trends toward the increasing influence of

[68] Bijker notes, however, that the STS constructivist analysis is politically neutral in that it does not necessarily work to the benefit of less powerful or victimized groups.

[69] Ecological modernization challenges the contentions of environmentalists (and radical eco-centrists, treadmill-of-production, demodernization, and deindustrialization theorists) that a fundamental reorganization of modern society – its industrialized production system, capitalism, and centralized states – is necessary to this goal. EM does acknowledge the need for some fundamental structural remedies of "design faults" in modern industrial production and consumption [Mol 2000]. However, EM advocates an ecological rationality that is independent of and on equal footing with others such as the prevailing economic rationality [Mol 1995].

consumers in the (re)organization of production and consumption chains. Privatization and liberalization in Dutch utility markets, for example, are expanding or differentiating consumers' relationships towards providers from traditional captive consumer to customer, citizen-consumer, and participant or co-provider. Citizen-consumers, "conscious citizens who may take individual action to serve social or environmental goals" [van Vliet 2002] are increasingly catered to by new marketing schemes like green-electricity and a variety of eco- and fair-trade product labeling. Their opinions seem to matter more than ever for business and policy makers. "Consumer-oriented monitoring" of commercial service providers and corporate producers "can also allow for counter-surveillance and strengthen the potentials of citizen-consumers to change the infrastructures of consumption" towards lower or improved use of resources [van den Burg 2001]. This is best combined with consumers' self-monitoring of their own resource and energy use. Both ends of the discretion continuum, or to use Renn's terminology, both types of risk communication (primary and secondary) for environmental improvement, are then covered. The interview-version of the ECO2 energy software was designed to permit both types of consumer-oriented monitoring.

Anyone trying to put these theories into practice should be mindful of certain potential pitfalls. First there are the lessons from cultural theories of power. Section 2.3.1 described the SCOT accounts of increasing "obduracy" and structural power as technological regimes progressively stabilize. Cultural theory says that social opportunities for the use of (scientific) knowledge are not homogeneous but depend on people's social positions, especially those of power and dependency, which culturally "inscribe" for people what they are permitted to "know" and how they are permitted to act. These boundaries are often implicit and "culturally rehearsed" [Irwin 1996b]. Social and institutional resistance to change also extends to actors' knowledge of, and negotiability of their role in sustaining, given constraints. Scientific knowledge that is socially legitimate and practically useful to people must be sensitive to these social and epistemological points: It must be "reflexive and self-aware" [Irwin 1996b]. This is true for the scientific message of the Energy-Revealing approach, and it is likely true of the more sociological message of the Social-Revealing approach as well. We must heed the caveats concerning psychological, social, and cultural variability and nuances of knowledge and communication discussed throughout this chapter, and not expect blanket applicability of this or any other risk communication approach.

Second, the preceding account of the development of risk communication described the shift in emphasis towards the receivers' active role in the communications process, the increasing importance of this process relative to "product" or message, and the recognition of the diversity of perceptions and values across the spectrum of groups and cultures that comprise the audiences or recipients. On the one hand, this conforms well to the participatory communications mode and its goals of furthering and deepening democratic participation, to which the application suggested here lends itself nicely. On the other hand, if the messages of both visibility approaches are sufficiently diluted or subordinated to process, or if the equal importance of all levels of "local," "contextual," or alternative knowledge among audience groups is insisted on (the vogue among some cultural reconstructionists looking into public knowledge), then the point of this research may be lost. A two-way open exchange and participation are necessary for legitimizing process, but it should not be championed at the cost of either the complex realities of environmental science, processes, and risks or those of societal drivers, interaction, and causality which the two approaches treat. This tension between breadth, subtlety, diversity, and cultural (disciplinary) inclusiveness on the one hand and research aims on the other is likely common to many trans-disciplinary projects. A suitable balance must be struck.

4. FINAL COMMENTS AND FURTHER APPLICATIONS

This section offers some final comments on the chapter's themes and points towards the empirical application of Chapter 5.

The Social-Revealing and broadened social science parent perspectives assert that "understanding the dynamics of energy demand (or demand for the services energy makes possible) is an exercise in understanding socio-technical change and the co-evolution of infrastructures, devices, routines and habits" [Wilhite 2000]. Analysis of both technological and concurrent societal development is thus necessary. This sort of analysis constitutes the third tier of energy consumption research: not (1) energy per se, or (2) just energy services, but (3) energy-related practices, sometimes addressed as a form and part of general consumption. Following in this vein, indicators of non-discretionary energy consumption might gauge end-users' perceptions and help improve the accuracy of perceptions of "the networks and infrastructures that together shape possibilities and choices that then form the subject of social negotiation" [Shove 1998]. Third tier-type research is at the cutting edge and is rarely done; bringing this perspective to the lay end-user is even rarer.

This type of treatment of energy consumption departs from conventions when, for instance, it treats demand as malleable in order to posit stemming the continuous escalation in energy demand. But in fact, laypeople may be more receptive to the concept of malleable demand, which harmonizes with common "folk" sense, than analysts or policy makers towing the line of political correctness or beholden to parties who benefit from demand escalation.

Bottom-up policy and market pull, secondary effect of risk communication, has been suggested as a desirable product of communication about non-discretionary energy consumption. Focusing on the social-institutional/behavioral boundary with respect to sustainable energy consumption – and encapsulating this in the proper communications framework – could influence the public's willingness and desire to support environmentally favorable technological, social, and institutional changes.[70]

It has been suggested that social norms often change according to the evolution of technological systems. Establishing the direction of causality is important, since decision makers responsible for technological parameters can be much more easily identified and potentially influenced than those accountable for social norms, if they exist at all. However, enhancing collective socio-technical awareness is also desirable.

People, including policy-makers, are certainly limited in their capacity for "independent action" vis-à-vis large-scale, entrenched socio-technical systems. And their influence may be felt only at certain moments or times of change [Shove 1998]. Still, such moments seem more common with the increasing pace of change today both in technology and society, as well as the increasing availability of information about these changes. The current socio-technological order is not yet too advanced in its evolution for beneficial shaping of its characteristics, nor are all of its new social norms inexorably entrenched.

[70] At some point, greater public insight may be politically counter-productive. "At a deeper level of global analysis ... a much richer data-bank of information which is of an economic, demographic, and socio-economic-technological interactive nature ... {might} then reveal much {greater} challenges with regard to the need of policy changes, lifestyle changes, industrial, and commercial changes ... and fundamental changes in the mechanisms and framework with presently guide and control the global environment" [Whiston 2000]. The resistance to such deep examination and radical changes would be huge and seems to doom any (democratic) attempt to effect them.

Chapter 4

FIELD STUDY WITH COMPUTER-AIDED INTERVIEWS

Change the environment; do not try to change man.
Design Science, 1969

All is foreseen, but freedom of choice is given.
Ethics of the Fathers, 3:19

1. INTRODUCTION

The preceding chapter examined lay knowledge, perceptions of, and risk communication concerning, less discretionary influences on energy consumption, in particular their nature, extent, variation, and pliability. This chapter explores how the voluntaristic/structural (e.g. behavioral/ institutional) division in energy consumption can be captured in a communicative tool for laypeople that could enhance their ability or desire to support environmentally favorable individual, technological, social, or institutional measures. Cutting-edge integrated assessment-based models are a logical starting point, since they expose some of the deeper social, institutional, and technological connections to energy and environmental trends.[71] Would a reconfiguration of a pre-existing personal and regional

[71] Whiston calls indicators that expose the deeper social-institutional connections to environmental states and trends "second-order environmental indicators." These connections are best shown dynamically and interactively [Whiston 2000]. Hypothesized third-order indicators "relat[e] to values, laws, legislation (and their influence, effectiveness, failure points, obstacles and barriers ... and the underlying rate of change," and thus seem in some ways enhanced "response" indicators from the Pressure-State-

energy calculator for Switzerland, a software program based on such a model, be useful in pointing in this direction, at least as a springboard for discussions and questions in interviews or focus groups?

Gregor Dürrenberger and Christoph Hartmann's original ECO_2 Calculator program[72] incorporated both household and Swiss national levels but largely separated them into "personal" and national (or regional) modules. Working together over a period of months to produce interview version 2.0 of the program,[73] we combined elements from the two modules in such a way as to allow the user to pose and at least partly answer the questions described below in the next section. The redesigned version would bring less-discretionary elements from the regional model, originally designed primarily for policy makers and planners, into a close interaction with the end-users' familiar household settings. In order to answer the set of user questions and other descriptive questions more fully, to test hypotheses, and to gauge such a program's usefulness as a tool for communication and education, it was clear the final program should be tested in an interview or focus group setting. Although focus groups would have added an element of social dynamism, the novelty and pioneering character of this effort, as well as constraints on time and manpower, argued for beginning with a set of structured, pilot interview sessions.

2. EXPERIMENTAL HYPOTHESES AND USER QUESTIONS

The following is a presentation of starting research hypotheses and brief annotations on them. These hypotheses generally served to frame the field study, strictly or loosely depending on the hypothesis. However, a number of other associated themes were also explored and are reported on in the results section. In the shaded box after the hypotheses is a list of related questions formulated from the perspective of a hypothetical user of the software. These questions steered the restructuring of the original ECO_2 software application into the "interview version" (that is, we redesigned and combined modules in such a way as to enable the user to explore these questions). In addition,

---

Response environmental indicator typology. The interviews described here explored the acceptability, utility, and workability of second-order type indicators for lay users.

72 A recent commercial English version of the ECO_2 calculators can be found at http://www.ecospeed.ch/ie/e/privat.html Other Internet versions are accessible at http://www.novatlantis.ch/eco2_pers/ and http://www.novatlantis.ch/eco2_regio/.

73 Gregor Dürrenberger and Christoph Hartmann, Der persönliche ECO_2 Rechner, Interview-version 2.0, 2001, ETH Zürich.

the issues and functions they describe were raised and addressed in the course of the interviewees' interactive use of the software during the interviews.

The first few hypotheses are more like philosophical musings than empirically testable hypotheses, but they may be useful in showing more of the thought process behind the development of the software and interviews.

A. **Freedom of choice:** Everyone has some control over a (non-trivial) portion of their energy use, even within the constraints of any particular "lifestyle" group.

This introductory framing hypothesis challenges the notion prevalent in some sociological schools of thought that nearly all (in this case energy-relevant) behavior is structurally determined. Yet, it suggests that part of one's energy consumption is also non-discretionary.

As suggested in Chapter 3, at the societal level, general or universal constraints found at the extreme high-end of the energy discretion continuum can be considered almost wholly non-discretionary (see Figure 1-3 in Chapter 1). These may apply to as yet technologically impossible choices or ones unavailable in a given place. They also apply to social norms and related antecedent technology choices – like historical trajectories for technology and infrastructure (e.g. lock-ins) – that highly shape the choice sets and lifestyles currently prevalent in society. Strongly individualistic or technically adept people might overcome some of these constraints, but most must conform to some extent to their operation in social and professional circles.

B. People can distinguish between those aspects of consumption they can directly control and aspects they cannot.

This adds a personal distinction between discretionary and less discretionary. Falsifying findings could be either individuals who are unable to make the distinction when by all accounts it exists; or those who feel all aspects are either one or the other; i.e. that they have either zero or unlimited degrees of freedom.

Capability of distinguishing is separate from exercising one's discretionary powers (specifically, in a direction of restraint) or, to put it in ethical terms, recognition of discretion does not necessarily imply a willingness to embrace the responsibility inherent in the freedom of choice.

C. (Non)Discretionary accounting: People relatively open to communication about energy use in society and interested in examining their own individual energy consumption benefit from a greater clarification, and personal assessment, of which determinants of their consumption they can change easily on their own (discretionary) and which parts they have less direct influence over (less discretionary); and the reasons for this categorization.

D. Perception of less-discretionary influences: It may be useful (for energy conservation purposes) to further the accurate understanding of this distinction, i.e. to replace perceptions of the degree of discretion with a more accurate rendering of the actual distinction in specific cases like housing, transportation, or leisure.

These basic hypotheses assume a certain level of motivation and engagement with the issues, which was true of most of the subjects recruited for the interviews. (The problem of higher engagement among the sample set than the general population is common to this sort of research experiment.)

Two components mentioned above are clearly differentiated here, a general or universal clarification of non-discretionary factors (achieved in the ECO_2 software module involving technological, demographic, and social factors), and a separate personal assessment (achieved in the household module). The latter involves both more clearly discretionary direct variables as well as personal "situational" non-discretionary constraints.

E. Communicating about energy consumption: Informing householders about structural, technological, economic, and social constraints on their household consumption and energy usage – along with information specific to the household – helps further sustainable energy consumption more than focusing on behavioral change alone.

This hypothesis is at the heart of the research outlook and effort. Risk communication about energy consumption is useful insofar as it presents in some way the multiplicity – and unavoidably, complexity – of its determining factors. Simple, non-differentiated indicators are not always desirable for presenting highly complex problems to the public, especially when trying to portray potential "solutions" or intervention points.

The question of what is a proper measure of "effectiveness" of such communication is an especially difficult one and was not empirically resolved in the interviews. Positive results of this sort of risk communication could consist of increased levels of awareness, behavioral intention, behavioral changes in household consumption, and/or changes in consumer or political action. But we could not necessarily expect any of these results to follow from participation in a single interview session, no matter how compelling or in-depth. Nor could we rely on interview follow-up questions regarding participants' intentions to take action. The best we could do was to extract this measure indirectly through various techniques in the interview, mainly by trying to assess the degree of learning the subjects experienced in the course of the interview sessions.

Hypotheses F through H largely relate to the better construction of energy risk communication tools like ECO_2. They are more specific and technical than the preceding hypotheses. Again, they take (existing) information-communication approaches as a starting point and, following the orientation of the hypotheses A through E, posit certain improvements.

F. **Aggregate data comparisons:** It is useful and instructive to present comparisons of a person's energy usage, in total and across sub-categories, to the national average and to that of other members of her household or peer group.

Aggregate data comparisons were already facilitated to some extent in the earlier version of the ECO_2 software. An extension of this idea (realized in the newer version) makes explicit links between the individual and the aggregate level, especially in helping visualize a flow from the former to the latter, by allowing the user to answer question 8 in the shaded box below. (This is the reverse of the effect of movement from the broad-scale level to the individual in question 5, for example.)

G. **Historical and cross-cultural comparisons:** Presentation of cross-cultural and/or inter-temporal comparisons is understand-able and useful (for stimulating thought on alternative socio-technical pathways). That is, identifying and elaborating on past political, economic, and social conditions as constraints on household energy use, and showing contemporary situations in other countries, is generally instructive.

H. Separation of the social from the technological:

1. Providing information about specifically social, cultural, and/or normative driving forces for energy consumption is important in energy communication.
2. Clear separation of technological from non-technological influences on household energy consumption is useful and important. That is, the user interface should separate specifically social, cultural, or normative driving forces for energy consumption from purely technological factors.

Questions the software was redesigned to allow the user to explore

1. Which direct household actions (e.g. behaviors, appliance purchases and use in heating, traveling, leisure time, etc.) do I think I could rather easily change? In what time frame? Which am I unable to change?
2. Of those I am able to change, which would I be willing to change in a direction of lower consumption; which not, and why?
3. What are the most important factors beyond my direct, immediate control constraining my energy use (in housing, travel, diet, etc.)?
4. What is the nature of the factors beyond my direct control? Technological? (e.g. prescribing or constraining availability of technologies, efficiency levels)? Demographic? Social (e.g. prescribing or proscribing types of consumption/activity levels)? Economic/institutional (e.g. establishing patterns of business and commerce that inflate the embodied energy of goods or setting incentives that strengthen the influence of other factors)?
5. How significant would be the effects of changes in general social or technological influences on my energy consumption in comparison with the effect of independent personal steps I might be willing to take? That is, how does the impact of "bottom-up" personal change compare with that of socio-technical change from the "top-down"?
6. What combinations of technological and social choices keep down future national energy use?
7. How easily and by which means could energy-relevant technological, social, or demographic forces be influenced, who are the decision makers involved (if any), and what could my role as an end-user be in this process?
8. What would national energy use be if everyone consumed as much as I do now or in the short- or mid-term? In combination with other broad-scale technological changes?
9. How do I react to/am I willing to confront non-technological influences on household energy use such as social, cultural, and normative factors?
10. Do I prefer a clear separation between technological and non-technological influences or a combined approach?
11. Do I find presentations of historical or cross-cultural comparisons of energy use helpful in illuminating our national energy situation?

The contrary, favored by the Social-Revealing approach, recommends that the interface not separate the two types of driving forces but rather show their interconnections and dependencies to the extent possible.

3. SOFTWARE DEVELOPMENT AND MODELING

This section benefited from material provided by Gregor Dürrenberger and Christoph Hartmann, the principal modelers and interface developers.

3.1 Original version of the Personal ECO2-Calculator

The first Personal ECO2 Calculator software tool was designed to provide users with a detailed accounting of how much energy they use in their daily lives. The program subdivides the individual user's energy balance (also called profile or budget) into 13 categories of everyday life: housing (or "living"), diet, private transportation (car), public transportation (train/bus), air travel, heating, miscellaneous consumption, and public services. All categories except the last three have both a direct and an embodied (grey) component (heating is only direct and miscellaneous consumption and public services are only grey).

As output, users can choose among any combination of the following two output variables:

1. Gross energy, end energy (gigajoules per year or watts)
2. Energy, CO_2 (tons per year)

As input, users set the values of the model variables and of selected technical parameters according to their personal situation. The output, the individual's energy balance or profile, is always visible on the interface, its categories are color-coded to match the color of the corresponding input fields on the interface, and it reflects the impact of every change in input in real-time.

3.2 Interview version of the Personal ECO_2-Calculator

The ECO_2-interview version tunes the Personal ECO_2-Calculator according to the needs of this particular study. It does not differentiate between output variables: All results are given in (gross) energy units (gigajoules per year or watts). The interview version consists of the original Personal ECO_2-Calculator and extended features, as well as an additional long-term screen with individual and Swiss aggregate (national) energy displays.

As an extension of the original Personal ECO_2-Calculator, the ECO_2-interview version allows users to compare their current energy consumption with a series of possible alternative (in the interview context, conservation) consumption patterns. The program permits users to generate and save alternative profiles, corresponding to different short and mid-term behaviors and purchases, in addition to their current (status quo) profiles.

On the separate long-term screen (Figure 4-1 and Figure 4-2) unique to the interview-version, long-term social and technological changes can be specified in order to account for less personally discretionary impacts on individuals' profiles. Apart from a baseline assumption reflecting projected trends in social and technological development out to the year 2030 (based on data from the Swiss Federal Office of Energy), users can define individual projections by varying 14 social and technical parameters. Of these, six technological parameters (energy efficiencies in industry, household, and mobility sectors) – often referred to for short as "Type I" parameters – have been modeled to show a direct impact on the user's individual profile, while eight social and demographic parameters (development of demography, household size, per capita use of floor space, and various mobility trends) – often referred to as "Type II" – have no impact on the individual's profile but are relevant for overall Swiss energy requirements (i.e. their impacts register on Swiss national and average energy budgets also displayed on the long-term screen). Each parameter can be set at five different levels: 2030 trend level, two levels above, and two below. The program allows users to run through the options twice for each time-horizon and saves all four data sets for later analysis. Table 4-1 lays out all of these parameters and the five possible values each can take. (Percentage values refer to the percent improvement from approximately current (year 2000) values. The middle value (third from the top) for each parameter is always a trend value projection for the year 2030. Everything is on a yearly basis.)

Figure 4-1 shows the original graphical interface (in German) for the entire long-term screen with all output patterns. Various output graphs are displayed on the left and settable parameters (as well as some other program function buttons) on the right. The left-hand output display is divided into two parts. The upper part shows two sets of personal energy balances with four bars each, while the lower part exhibits data for aggregate Swiss national energy consumption.

Table 4-1. Selectable values for Type I and Type II parameters

TYPE I PARAMETERS		TYPE II PARAMETERS	
Manufacturing efficiency (avg)	20%	**Modal Split Persons**	10%
	15%		20%
	12%		25%
	10%		30%
	7%		40%
Residential heating, cooling, A/C	55%	**General consumption**	10%
	45%		20%
	35%		25%
	25%		30%
	15%		40%
Efficiency of goods transportation	0%	**Population**	7,850,000
	1.5%		7,650,000
	3%		7,450,000
	6%		7,250,000
	15%		7,050,000
Modal split goods	30%	**Automobile travel**	3500 km (1970)
	40%		8000 km
	45%		11000 km
	50%		13000 km
	60%		16000 km (current USA)
Electricity mix	Current Swiss mix	**Auto occupancy (avg)**	1.4 people
	European mix		1.5 people
	Hydropower		1.6 people
	AKW (Swiss electr. co. avg. efficiency)		1.7 people
	EWZ (Swiss electr. co. avg. efficiency)		1.8 people
	Green electricity mix	**Air travel avg.**	200 km
	Photovoltaic		300 km
	Wind		400 km
	Brown coal		600 km
			2200 km (USA)

TYPE I PARAMETERS		TYPE II PARAMETERS	
		Residential living space/person	40 sq. meters
			47 sq. meters
			54 sq. meters
			61 sq. meters
			68 sq. meters

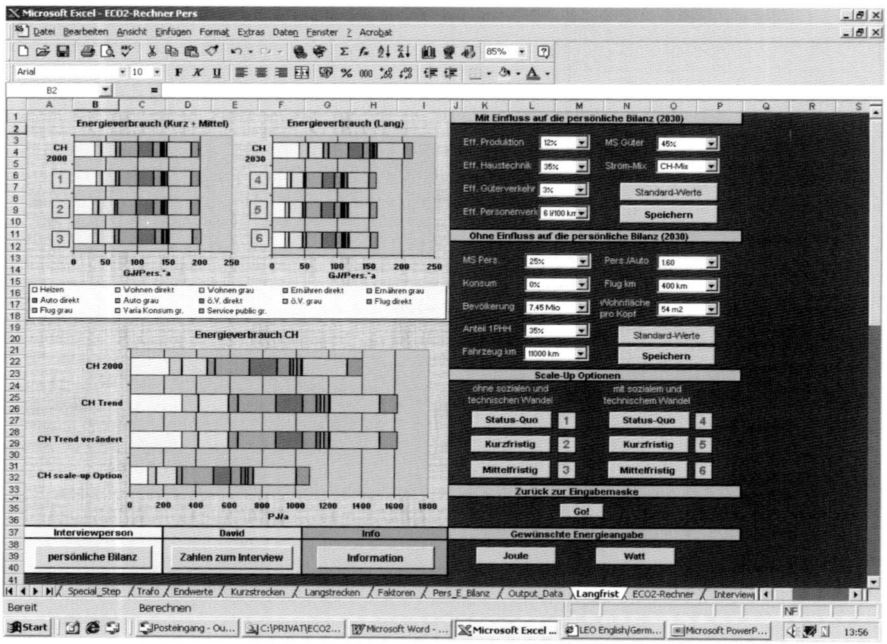

Figure 4-1. Long-term (non-discretionary) screen

The left set of the displayed personal energy balances comprise, from top to bottom, the Swiss per capita average in 2000 and the user's three personal energy balances generated in the previous (personal household) ECO_2 screen, i.e. the user's status quo, short-term, and mid-term conservation profiles. The right-hand set of personal energy balances consists of the matching graphs on the left under the hypothetical influence of (tuned to) the technological parameters at levels set by the program user (far right). A close-up translated snapshot of these eight graphs from a typical interview session is offered in Figure 4-2 below.

The lower output section in Figure 4-1 shows various energy profiles for the whole of Switzerland (in petajoules or gigawatts). The first bar shows Swiss energy consumption in 2000, subdivided according to the same categories as in the individual profiles but aggregated. The second bar shows the projected trend values for 2030. The third bar shows Swiss energy consumption in 2030 according to the parameters set by the user. The final part displays results from applying the scale-up function by means of buttons on the lower right. This last function is designed as a long-term thought-experiment: How much energy would be consumed in Switzerland if everyone's consumption behavior resembled the user's? Using the buttons on the lower right, each of the individual energy balances, i.e. status quo, short-term, and mid-term, can be scaled-up with or without the assumption of concomitant long-term social and technological changes.

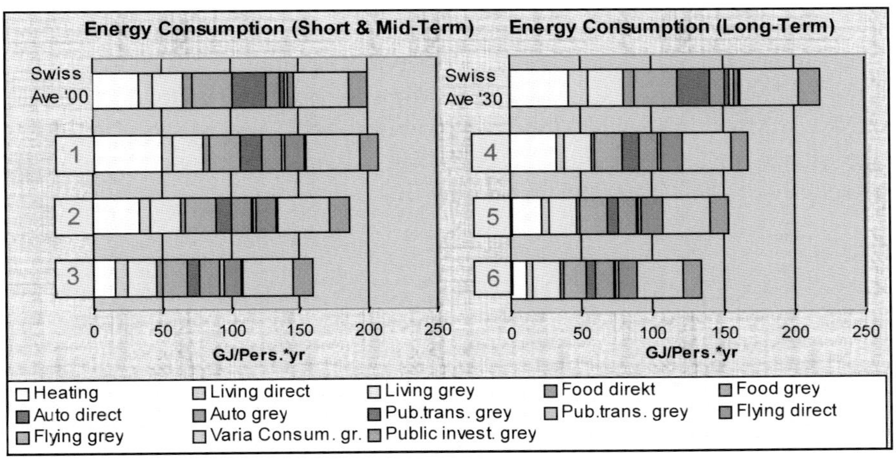

Figure 4-2. Close-up snapshot of individual user output from long-term screen, with key for color-coded categories.

Output from the program can be inspected visually on the screen or on printouts, and Excel data generated during a session can be analyzed later using typical Excel and statistical functions and/or can be exported to a statistical or database program.

For further technical details on the modeling and software, see [Goldblatt 2005].

4. INTERVIEWEE RECRUITMENT

Over one hundred people were recruited, of which twenty-one were selected to go through the full interview sessions. They ranged in age, income, education, profession (or students' field of study), housing, car use, environmental leanings, and other variables. We did not make a random sampling of subjects but rather selected a subject pool that represented a range of characteristics. We decided against sampling randomly since, due to the length and complexity of the interviews and constraints on time and manpower, we did not deem it feasible to conduct a sufficient number of interviews to reach a statistically generalizable sample size. One option that could have possibly allowed statistical inferences to be drawn for the entire population, interviewing only university students, was rejected as too limiting for the purposes of a pilot study experiment and overly tedious for the interviewer.[74] Close to half of the subjects were university students (who were generally more available and in need of the compensation offered, 100 Swiss francs for the session). Student interview sessions lasted an average of three and one-half hours; most interviews were done in single days, with a break at the midpoint. Later interviews with businesspeople were streamlined to approximately three hours. All interviews were conducted personally by the author, the majority in English.

5. INTERVIEW DESCRIPTION AND GUIDELINE

1. In a set of short preliminary questions, the subject's initial opinions and impressions are assessed, including her degree of engagement with energy problems and environmental issues, her self-ranking as an energy consumer, her most energy-intensive activities, faith in technology or behavior change as solutions, and so forth.
2. After various other preliminaries, with help from the interviewer, the subject enters information about her housing, heating, transportation, diet, miscellaneous consumption, and so on (most both direct and embodied) to generate her status quo energy consumption profile in gigajoules or watts per year. She compares her profile to the average and to one for her household type, in total and across categories.[75] She is

[74]This statement is much more a reflection of the author's limited tolerance for tedium than of any tedious tendencies among Swiss students.

[75] The categories are listed below in section 8.2 "Terms and comments."

asked to try to account for the size of the most surprising or largest category in her profile.

3. The subject then goes through the entered activities again and categorizes them according to whether or not she is able to change them in a direction of lower energy consumption, and if so, whether and to what extent she is willing.[76] The subject quantifies these "able and willing" categories with new choices on the user interface; the program saves these and generates new profiles (one or two, representing short-and mid-term conservation efforts). The subject compares them to her status quo profile.
4. Moving to the screen with long-term, national non-personally discretionary factors (Figure 4-1), the subject makes choices in future (or simply hypothetical present) levels of various technological variables or parameters ("Type I" variables), including efficiencies of industrial production, residential and commercial heating/air conditioning/ventilation, and personal and goods transportation; modal split for goods transportation; and electric power generation mix. Each variable has a 2030 trend level and four other levels arrayed around the trend. The effect of implementing these levels produces changes that are shown in corresponding (2030) personal profiles displayed to the right of the status quo, short-, and mid-term profiles (as well as in the future Swiss average and aggregate graphs). The subject is asked to make a series of comparisons of the effects on the profiles of various levels and combinations of these parameters. The subject is also asked to compare her conservation profile(s) with her status quo profile under the influence of technological changes in order to see in which circumstances such changes are more or less important for reducing her energy consumption than personally initiated changes alone (as reflected in the conservation profiles; i.e. the subject compares graph 4 with graph 3 (or 2) in Figure 4-2, considering total energy, separate activity categories, or both). Technology variables are maximized or minimized, alone or in combinations, to simulate a variety of extreme circumstances; the comparison questions are repeated, and depending on the person, some of the conclusions change as thresholds are crossed.
5. The subject now chooses levels for demographic and social variables in Switzerland ("Type II" variables), including population, percentage of single-person households, modal split for personal travel, general

[76] The program can naturally also be used in a predictive fashion to show the energy effects of expected future changes, whether increases or decreases. For the sake of illustration and comparison, we quantified only potential downward changes.

material consumption, average driving and flying levels, auto occupancy rates, and per capita housing floor space. Unlike the technological parameters, changes in these variables are not modeled to affect changes in the subject's personal energy profiles, so in order to see their effects, the subject is directed to focus on the Swiss average (top left of the screen shown in Figure 4-1) and/or Swiss aggregate displays (bottom left).

6. Once everything is set, the subject is asked to assess the effect of all of the combined variables on (future) Swiss energy use, to see which are the most important determinants (and especially whether technological or social predominate); and to muse on the connections between the two classes of variables, here artificially separated.
7. The highest and lowest values for driving and flying are pre-programmed to current US levels and Swiss levels in 1970, respectively. As cross-cultural or cross-temporal illustrations and social thought experiments, the interviewer selects these and asks for the subject's hypothetical personal reactions to a Switzerland in which these levels obtained.
8. As a section follow-up, the interviewer asks whether the interviewee's perceptions have changed regarding the most important determinants of energy consumption, on a personal or national level, and whether she would now rank herself differently in terms of her relative consumption level. What is the subject's view of the relative impact of (personal) discretionary vs. non-discretionary factors now that she has seen them both modeled?
9. To contextualize for a moment, how absolutely "non-discretionary" does she consider the modeled technological and social variables? For a given activity sector like heating or transportation, how easily does she think the relevant (modeled) variables could be changed to favorable levels; and what are possible intervention points and pathways for end-users as well as for policy makers?
10. The final function scales up any of the subject's personal profiles to Swiss national levels to answer the question: "What would national energy use be if everyone consumed the way I do, currently or with conservation measures applied, and with or without blanket technological changes?" (question 8 in the box of user questions, section 2). How does the result compare with present or future (trend) levels in Switzerland (as appropriate) and thus how does the subject's lifestyle compare to others'? How do the subject's scaled-up profiles, even the personal conservation profiles combined with advanced technology, compare to ecological energy thresholds like the 2000 Watt per capita society? What additional, unaddressed parts of the story does this suggest?

11. The interview concludes with short debriefing questions to address some final substantive points and solicit the subject's assessment of the program and the interview experience.

6. DATA CAPTURE, STORAGE, AND ANALYSIS MEANS

During the interviews, data were recorded through a combination of selective audio cassette recording, pre-prepared worksheets filled out by the interviewer or the subject, on-site hand-written notes, and Excel data generated by the software and saved for each subject at the interview's conclusion.

Portions of the audio recordings were transcribed or their data otherwise extracted into a Zoot 3.1[77] program database. Zoot and subsequently Excel 2000 served as the main data storage, manipulation, and analysis tools.

7. SUBJECTS' BIOGRAPHICAL PROFILES

See Table 4-2 for profiles of the 21 subjects.

Note: For anonymous reporting purposes, subjects were assigned subject numbers and will be referred to in the text by "S" followed by their number, e.g. S8.

8. RESULTS

Note to the reader: The rest of this chapter presents a selection of results generated by the interviews and subsequent data analysis. Although abbreviated, they are described in a quantity and at a level of technical detail judged sufficient to satisfy an interested researcher. This may prove too much for some readers. For their benefit, we have provided summary boxes at intervals throughout the text. Also, the presentation in the latter half of this section is somewhat less technical and more descriptive and reflective.

[77] http://www.zootsoftware.com/

Table 4-2. Subjects' biographical profiles

Gender	16 male, 5 female
Age	Avg. (mean) 32.3 years, s.d. (standard deviation) 11.7 years
Residence	Zurich (16), Aarau (1), Brüttisellen (1), Trimmis (1), Lausanne (1), Prilly (1)
Educational Level	University (undergraduate or higher) 18; trade-school or equivalent: 3.
Profession	13 7th / 8th semester ETH undergraduates, doctoral students, or recent graduates:78 8 professionals: private banker (2), bank worker (1), bank director (1), commodity trader (1), business owner (1), biomedical scientist/ entrepreneur (1), clergyman (1)
Household Size	Avg. 2.70, s.d. 1.08
Household Income (Monthly)	Avg. Swiss fr. 7994.74, s.d. 3516.70
Personal Income (Monthly)[79]	Avg. Swiss fr. 3235.00, s.d. 2251.34
Housing	18 apartments, one double-apt., two houses
Auto Use (Car or Motorcycle)	10 autos, 2 motorcycles (with autos)
Environmentalism[80]	Avg. 0.79 (neutral to somewhat green), s.d. 1.04
Engagement with Energy Issues	Avg. 67.5%, s.d. 29.36%

In addition, some readers may find the shift too sudden and swift from the sweeping societal perspective of much of the previous chapters to a level of reporting on experimental subjects' personal views and reactions. If so, we ask for patience until the final chapter where these micro-experimental results are explicitly tied into the larger theory and discussion of the broader book.

78 Students' fields of study: physics (1), food sciences (2), environmental engineering (4), electrical engineering (1), survey engineering (1), biology (1), computer science (1), architecture (1), process engineering (doctoral- 1).

79 Household income/household size

80 Numerical interpretation: 2: very green (environmentalist); 1: somewhat green; 0: neutral; -1: somewhat anti-green; -2: very anti-green.

8.1 Subject groups

The three groups will be introduced now because much of the further description and analysis makes reference to them.

The subjects were categorized into three groups on the basis of strong patterns suggested by data from the first part of the interview – through the subjects' encounter and experimentation with technological variables (through step 4 in the interview description above). Up to this point, subjects had formed first impressions of the relative impacts of technological changes compared to those of personal conservation measures, based on their own idiosyncratic energy profiles and considered before the interviews introduced the national-level social and demographic parameters. I made a short, dense textual summary of all of these data for each subject. I then discerned the three groups by "eyeing" these summaries all at once. (More detail is given in 8.4.1 below.)

Subsequent data were analyzed with reference to these three groups and to the data as a whole. For a given hypothesis or theme, notable group patterns, or their absence, were described. Often strong group consistency was maintained, which meant that the groups generally maintained their identity across the range of themes in the data. However, these groups are not the only ones that could be formed on the basis of these data, nor, as will be shown, do all the data conform to them.

In the course of the results presentation and discussion we will see to what extent these analytical functional groups hold up as differentiated lifestyle groups of householders some of whom "evolve a variety of approaches to (sustainable) consumption and (co)provision" [Spaargaren 2003].

8.1.1 Group I data summary (through interview step 4)

(Students, under-average energy consumers, personal/technological "balanced")

(Subjects S1, S2, S3,S4,S8,S9,S10,S12,S13,S14,S16,S18,S20, and S21.)

Group I comprises two-thirds of the subjects. Members are mostly students or recent students who still lead something of a "student" lifestyle (even if now working). They are mostly under-average energy consumers (average deviation from Swiss national average: -30.48 gigajoules (GJ)/person*yr, standard deviation: 26.50 GJ/person*yr.[81]). They display a

[81] Σ_a^m Status Quo energy consumption - Σ_a^m Swiss_average, where a: Heating; b: Housing direct; c: Housing grey; d: Diet direct; e: Diet grey; f: Private transportation (Auto)

modest ability and willingness to reduce in the mid-term by an average of 23.02 GJ/person*yr, with a standard deviation of: 15.21 GJ/person*yr.[82] On the question of which has a greater downward effect on their energy consumption, blanket technological or personal interventions, Group I members generally see a balance between technological and personal, often with personal dominant for Flying and/or Auto and Heating. This balance is usually at least somewhat sensitive to the technological levels reached.

Two past Swiss studies on energy use and the public

In a preliminary research phase, the author canvassed past Swiss studies for relevant data. Two prior studies of Swiss householders' views on energy stood out, one from the 1980s and the other from the 1990s. Energy in Everyday Life (L'Energie au Quotidien) [Bovay 1987] is a summary of findings from 56 in-depth interviews with French-speaking Swiss on the subject of energy, conducted in the 1980s as a study within the Swiss Research Project (NFP) 44. The larger section consists of sociological research into households' energy consumption involved in daily activities like lighting, heating, water use, recycling, and transportation. Urs Dahinden's Democratizing Environmental Policy (Demokratisierung der Umweltpolitik) [Dahinden 2000] asks how public input can be combined with expert knowledge to further a "democratization" of environmental policy. His empirical case study is based on extensive focus group discussions of the role of economic instruments in energy policy, among them information for consumers and energy specialists. Themes common to both studies include personal vs. state-corporate responsibility; lay views on the role of information; the evolution of comfort norms and demand escalation; and social shaping possibilities. The author's own investigations and experiments, discussed in this chapter, partly elaborate on and partly go beyond these themes.

8.1.2 Group II data summary

(Younger professionals, above-average energy consumers, personal/ technological "balanced")

(Subjects S5, S6, S15, S19.)

Much above-average energy users (average: 144.20 GJ/ person*yr, standard deviation: 74.89) in single- or two-member households (although one, S15, was expecting a child and therefore anticipated some lifestyle changes, and one was a middle-aged man whose children had moved out but

direct; g: Private transportation (Auto) grey; h: Public transportation direct; i: Public transportation grey; j: Air travel (Flying) direct; k: Air travel (Flying) grey; l: Miscellaneous consumption grey; m: Public investment grey (See 8.2 "Terms and comments" below).

[82] Σ_a^m StatusQuo1(E_value) - Σ_a^m Could_reduce (See 8.2 "Terms and definitions" below).

who remained in the same large apartment), willing to make many changes in the short or more typically mid-term, which may have strong impacts on their energy profiles (average reduction: 77.62, standard deviation: 69.40). They see some balance between personal and technological or at least personal dominance for a few categories (typically at least Flying).

8.1.3 Group III data summary

(Family professionals, average energy consumers, resistant to personal change)

(Subjects S7, S11, S17.)

Average or somewhat above-average energy users (average: 21.59 GJ/person*yr, standard deviation: 27.03), all with larger (four) member households, resistant to any change in personal behavior or circumstances (minimal changes and/or energy impacts; average reduction: 7.12, standard deviation: 1.59). They recognize that their personal energy profiles are affected almost exclusively by technological, rather than personal, changes (even though they later see that aggregate behavioral and demographic factors may have a strong influence on national consumption).

Table 4-3 shows a simple summary of some of the groups' main differentiating characteristics. (The last column, "μ Pers vs. Tech" (a dependent variable), conveys which type of intervention, personal or top-down technological measures, is most effective on average for that group, and is recognized as such.)

Table 4-3. Groups I, II, and III in simple contrast

	Age	Energy use	Household size	μ Pers vs. Tech
Group I	↓	↓	↑	= , Pers
Group II	–	↑	↓	Pers
Group III	↑	–	–↑	Tech

8.2 Terms and comments

As preliminaries, here are the English[83] terms for the energy sector subcategories that made up the energy profiles (both individual and national):

[83] The program user interface was in German.

a: Heating (direct only);
b: Housing (Living) direct;
c: Housing grey (i.e. embodied);
d: Diet direct (food storage and preparation);
e: Diet grey;
f: Private transportation (Auto) direct;
g: Private transportation (Auto) grey;
h: Public transportation direct;
i: Public transportation grey;
j: Air travel (Flying) direct[84];
k: Air travel (Flying) grey;
l: Miscellaneous consumption (grey);
m: Public investment (grey)

Numerical energy values are sometimes referred to using this labeling. For example, $StatusQuo_e$ refers to the energy value of the Diet grey activity (in gigajoules per year) in a user's status quo energy profile. $\Sigma(StatusQuo)_a^m$ refers to the sum of all the energy sector (i.e. activity type) values across the user's status quo profile, i.e. his total yearly energy use.

The short and mid-term conservation profiles generated in the first part of the interview (step 3) were referred to during the interview by the terms *can reduce* (and would) and *could reduce* (and would),[85] respectively. Changes the subjects identified as possible but personally undesirable (nuances explored only in the longer, student version of the interviews) were called *can reduce but don't want* to (in the short-term) and *could reduce but wouldn't want to*[86] (mid-term). Unchangeable items were sometimes labeled *cannot change* (and in the streamlined interviews this latter category was the catch-all alternative to *could reduce (and would)*, implying any changes the subject could not and/or would not undertake). These terms are sometimes used in the short-hand formulas or discussion below.

[84] Air travel does not include business air travel, which is indirectly accounted for in the embodied energy of the other activity types. The air travel category therefore represents the energy used for personal or leisure purposes.

[85] *(Das) kann ich tun* and *könnte ich tun* in the German.

[86] *(Das) kann ich tun aber will nicht* and
könnte ich tun aber würde nicht

8.2.1 Note on correlation analysis

As part of the analysis, correlations were calculated between the quantitative data results (including $time_1/time_2$ assessments) and the variables age, household size, personal income, environmentalism, and energy-engagement. Correlations were calculated across all subjects (group correlations were not done). Where relevant, a summary of the moderate correlations found and some interpretations are provided. Moderate correlations are defined as those having correlation coefficients from +- 0.40 to +- 0.80. There were no strong correlations found (< -.80 or > .80)). Where no mention or comment is made, the correlation is weak (> -.40 or < 0.40).

The following presentation of results is organized loosely around the lettered hypotheses of section 2 and related themes. Within each section, units of data, sometimes corresponding to individual interview questions, are grouped and ordered for clarity.

8.3 Hypotheses A and B: Freedom of Choice and Capability

8.3.1 Total household energy reduced by means of mid-term conservation steps (inclusive of short-term steps)

Σ_a^m StatusQuo1 - Σ_a^m Could reduce

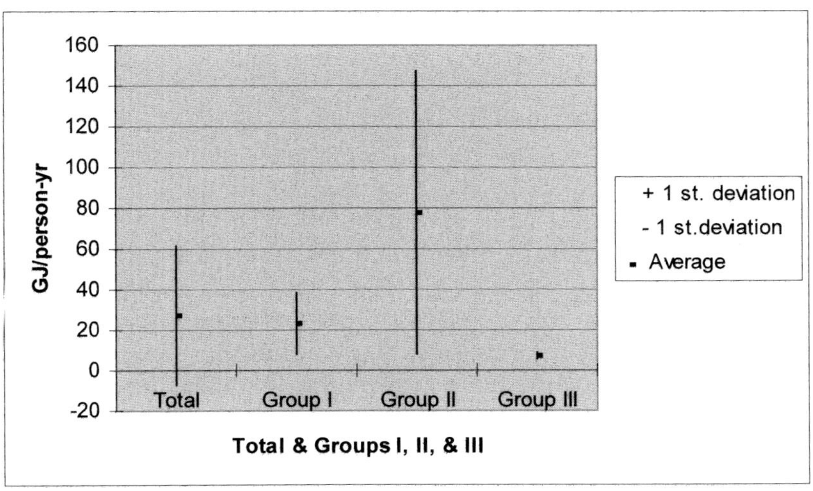

Figure 4-3. Total energy reduced from personal profile through short and mid-term conservation steps

As noted in the group data summaries, Group I is willing to make moderate quantitative energy reductions (close to the overall average), Group II more significant reductions, and Group III minimal ones (see Figure 4-3).

8.3.2 Correlations

Short-term energy use reductions and **personal income**: 0.6812. **Mid (and short-) term energy use reductions** and **personal income**: 0.5776. These two are again among the highest if not the highest correlations encountered in the data analysis. One interpretation is that those with higher incomes have more flexibility to reduce in the short and mid-term, in that their larger budgets offer some non-essential (financially discretionary) cushion in money and linked energy expenditure which can be reduced, and which subjects are willing to reduce to some extent. Several of the highest income earners from Group II acknowledged this when they differentiated between "need to have" and "nice to have" in their expenditures and activities. Restraint may occur at all levels of income, but the higher levels offer more opportunities for substitution and painless reduction. By contrast, a number of low income subjects from Group I saw themselves as living closer to the level of basic needs with minimal possibilities further to reduce energy expenditures. This aspect of *discretionary*, then, is quite similar to the economic notion of *discretionary income*.

Mid (and short-) term energy use reductions and **household size**: -0.423, a barely moderate negative correlation. As noted in Group III's data summary, members of this group, older family professionals with the consistently largest household size, are most resistant to changes from their status quo.

We **note** that in the subjects' viewing the energy profile and reduction possibilities as a whole across all household activities, it is possible to let reductions in one sector compensate for increases in another. If we permit the user to change the base information that includes household size (generally not done in the sessions), she can even "game" the program, for instance, by increasing future air travel but simulating a growth in family household size that reduces her personal (per capita) energy balance. These sorts of games, in which one sector is played off another, are possible when only the total energy profile is set as the target to minimize and not that of individual sectors. Such strategies fail if one reckons by sectoral energy (or overall CO_2 or climate-forcing effect).

8.4 Hypotheses C, D, and E: Non-discretionary accounting, Perception of less discretionary influences, and Communicating about Energy Consumption

8.4.1 Effects of technological change on subjects' energy use and the analytical extraction of Groups I-III

As described above in step 4 of the interview, subjects experimented with various (combinations of) levels of technological (Type I) long-term parameters and observed the modeled effects on their personal profiles, noting especially the impact on their status quo profile compared to what they could (and would) achieve by personal steps alone. This latter comparison was carried out when the first parameter, efficiency of production, was set (and the others remained at their default trend levels), and as a sensitivity test it was repeated when efficiency of production was maximized or minimized, after all Type I parameters had been set, and again after all Type I parameters had been maximized or minimized.

These experiments allowed subjects to form first impressions of the relative impacts of blanket Type I changes as compared to those of personal conservation measures, assessed of course only on the basis of their own idiosyncratic energy profiles and considered before Type II parameters were introduced on a national scale. The subjects' perceptions of the technological-personal comparison, as extracted from interview questions, were individually analyzed, but no total or group summary statistics were generated. Instead, a short, dense textual summary was made for each subject of these data and all of the prior data since the start of the interview. These were "eyed" and from them the three basic groups I, II, and III were discerned. Their construction and descriptions, as reported in the group summaries above (section 8.1), were based on these textual summaries.

8.4.1.1 Subjects' interim conclusions concerning the effects of technological change on their energy use

Step 4 of the interview concluded with a few overall open questions:

"What do {the preceding exercises} tell you about what's important to reduce your own energy consumption?"

"In general, what does this exercise alone suggest to you in terms of energy conservation policy?"

The main part of these results entered into the textual summaries that formed the basis of the categorization into Groups I, II, and III. A selection of additional salient comments from group members follows:

8.4.1.2 Group I

There were varying opinions in Group I on the *difficulty* of inducing personal or technical change – in addition to, and independent of one's stand on, the belabored issue of which is more *effective* in reducing energy consumption (this is explored further below, for example in 8.7). S2 found personal change generally more effective for her profiles, except at the highest efficiency levels, which she deemed unrealistic and in any case less easy to achieve than her own changes in the mid-term. But she still viewed personal change as harder to effect than normal levels of efficiency improvements, even though technological change faced significant political barriers. S10, by contrast, judged technological change to be more effective overall, but similarly found efforts to induce it more likely to bear fruit than trying to convince people to behave more "reasonably" (more "efficiently" per unit effort expended). Others refined this to suggest investing in the most (cost) effective Type I variables. For example, S3's experience with the software showed him that improvements in goods transportation efficiency do not seem to produce such dramatic reductions overall as equivalent improvements in transportation of people.

After completing the exercises with Type I variables alone, some believed in the effectiveness of a pure technological fix if the technological improvement were sufficiently high. S8, for example, at first concluded: "It's comfortable, too comfortable: I needn't do anything {but wait for technological development}," but later (after seeing the operation of Type II social and demographic variables on the aggregate) he lowered his confidence in Type I variables alone.

Among those of Group I who discovered that personal change was more effective and thought it was easier to achieve, some (e.g. S18) said they would put their newly heightened awareness into action in the household and assumed other similarly enlightened householders would do the same. However, other subjects (e.g. S12) said they would not substantially change their behavior as a result of seeing its significance in areas like transportation and flying demonstrated during the interview sessions.

S21 said she believed the processes of personal and technological change could reinforce each other. When citizens see the government taking an active role in stimulating technological improvement for environmental benefits, they are better prepared psychologically to do their part to conserve. Higher energy prices help reinforce this psychological effect, and independently they stimulate market forces to promote innovations in efficiencies.

Several subjects – including S8, S14, and S18 – spoke as if they assumed that findings true of their own profiles applied to everyone. S14, for example, was excited at discovering that a few relatively small personal changes (trivial compared to most of the efforts necessary for diffusing better technologies, he said) could dramatically improve his heating profile, and he wanted to conclude that this was generally the case. In such instances, it was pointed out to the subject that these properties were highly particular to the individual's profiles and not necessarily generalizable: We would soon examine the situation in the aggregate.

One subject in Group I (S18) differentiated technological improvements in categories according to the evenness of the distribution of their use or impact on the general population. Food and Living, he said, are activities done by everyone, so efficiency improvements here would be felt everywhere across the population. Flying, however, is still a more privileged activity, and technological progress in this area would not redound to the benefit of the whole population.

In applying this egalitarian reasoning, S18 implicitly took an approach more typical of environmental justice than a global CO_2 emissions perspective. Globally, it is likely that populations in less developed countries, the majority of whom never fly, will suffer greater physical or socio-economic damage from the long-term effects of climate change. And CO_2 released by aircraft at cruising altitudes evidently has a much greater climate-forcing potential than an equivalent quantity of CO_2 from ground-level activities. This argues for a comparison of technological improvements on the basis of the relative environmental mitigation they could provide, not distributional impact across user populations. In this particular case, a technical Energy- and Environment-Revealing calculation (in fact beyond what the ECO_2-interview version could provide) seems much more warranted than Social-Revealing reasoning.

8.4.1.3 Group II

In Group II, S15 concluded that personal steps were more important than any steps industry might take, since it is the individual who decides whether to place efficiency or environmental performance ahead of cost in her purchasing decision. However, efficiency improvements are necessary, and consumers must also (collectively) exert market pull to compel manufacturers to generate them. For her the direction of the impulse is clearly bottom-up: Manufacturers respond to expressed consumer demand,

and they would not try to impose highly efficient products on uninterested consumers.

S5 judged technological change and behavioral change to be roughly equal in effect but technological change to be more time-consuming and uncertain. Behavior is much more quickly amenable to change on an individual level, but the difficulty here is in the scale: it is much harder to motivate and reinforce across major portions of the population. Still, through the use of market instruments and, secondarily, possibly legal measures, it is still perhaps faster and surer than promoting technological change.

8.4.1.4 Group III

In Group III, S17 found technology dominant in effect but cautioned that the pace of development of energy-saving technologies was highly dependent on whether (price) incentives to save energy were in place.

8.4.2 BA4: Ability to influence personal energy consumption

One of the analytical features built into the interviews was a cluster of $time_1$/$time_2$ (t_1/t_2 or Before/After "BA") questions designed to measure the subjects' degree of learning or opinion change over the course of the interview. Some of the results of these measures are reported later in this chapter in section 8.8.1 "The program and interview as a communicative and educational tool: Subjects' degree of learning over the course of the session." Other results, however, relate to views on the relative impacts of technological (or other higher level) interventions and personal steps and are appropriate to discuss here.

The constituent answers in the Before/After questions were quantitative or were assigned ordinal rankings where possible, in order to allow for a numerical (often percentage) or quasi-numerical measure of change from $time_1$ to $time_2$. In this case, BA4: Ability to influence personal consumption, the average change for the total (all subjects together) was -3%, i.e. a slight downward revision in subjects' assessment of their ability to influence their personal energy consumption. Group I's average was -6%, twice the total average, suggesting the session brought home to some members of Group I the limits of their ability (or willingness, with which ability may be conflated in approaching this question) to influence their personal profiles. Groups II's average was +6%, twice the total average in the positive direction, but this was due to just one subject who saw a 25% greater personal influence by the end of the session. Each of Group III's members showed no change.

Definitional note for the Before/After (BA) questions

In trying to measure how much subjects learned from the interview and program on any particular point (variable α), we examined the difference between the state of the variable at $time_1$ and its state at $time_2$, that is, the paired difference d = variable α t_2 – variable α t_1. Thus, α's levels can be assigned any numerical value, as long as the same value scale is used at both t_1 and t_2. If there is no change, the paired difference is 0 (nil). For closed questions like the subject's self-ranking as an energy user (below), the variable can take on between 3 and 5 values, which when consistently assigned ordinal rankings, means that the most the paired difference can be is 4.

Note also that methodological problems with some interview questions, apparent only after the fact, made certain Before/After comparisons partially invalid and conclusions based on them therefore suspect. Only those comparisons that seem reliable and defensible are reported here.

8.4.3 BA5: Importance of behavior/personal influences on subjects' current energy use

The average for the total is -5%, i.e. an ex-post downward revision of their original assessment of personal influences on their current status quo energy use (probably in favor of technological or other factors) by about 5%. (This seems to conform to the total results for BA4 in section 8.4.2, possibly because the subjects' interpretation of the two sets of questions overlapped.) Group I's average was -7%, Group II's was +12.5% (but with a large standard deviation: 3 members had 0% and 1 member 50%), and Groups III's was -25% (but one subject's data were not available).

The total average of -5% suggests as a result of the interview session a tendency for a modest growth in recognition of the strength of the role of less discretionary factors (especially technological) in determining current energy consumption levels and a corresponding lowered weight given to personal factors. This is true on average of Group I members (-7%) while Group II members, as might be expected of a group of high-consuming professionals, showed no change or an increase in the weight given to personal factors. One member of Group III weighted technological factors much higher at the end (and personal factors even lower than a rather low rating initially) and so drove down the average for his small group.

8.4.4 BA6: $Time_2$ assessment of importance of identified influences on subjects' current energy use.

Near the beginning of the interview (step 2), subjects were asked to account for the size of the prominent sectors in their personal energy profiles. Near the end (step 8), they were asked how accurate their accounting had proved. The degree of change in their perceptions can be summarized by classifying subjects' experiences into one of four groups: a. No change – fully confirmed; b. Mostly confirmed, slightly changed; c. Partly confirmed, partly (i.e. assessment of some other determinants) more significantly changed; and d. Very significantly changed. The results are summarized in Figure 4-4 below and details for selected individual subjects presented next.

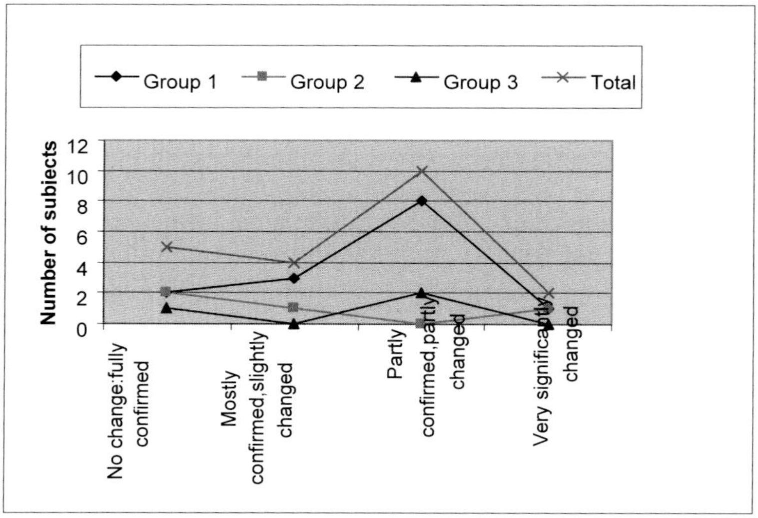

Figure 4-4. Change in perception of relative importance of influences on current household energy use

8.4.4.1 No change – fully confirmed

Total 5 members: Group I, 2 members; Group II, 2 members; Group III, 1 member.

Group I

S3: The importance of personal choice was largely confirmed generally and for air travel; for the embodied energy of food, personal choice now seemed somewhat less important.

8.4.4.2 Mostly confirmed, slightly changed

Total 4 members: Group I, 3 members; Group II, 1 member.

Group I

S21: This subject largely confirmed her initial assessments, but she did revise upward leisure activities in miscellaneous consumption. The subject said she would think about them more carefully in the future. On the other hand, at high technological efficiency levels, most categories were technology dominated, except air travel, which always remained personally dominated.

8.4.4.3 Partly confirmed, partly changed

Total 10 members: Group I, 8 members; Group III, 2 members.

Group I

S4 and S12 found flying more important than they realized previously, S10 saw no change, while S8 was surprised at his frequency of flying over the past ten years brought out by the exercise of recollecting trips. Many were surprised how large certain categories were, e.g. heating (S10, S8, S2), living (S10, S4), and food (S10, S8, S4 (confirmed)).

Group III

S17 had his assessment of the influence of personal factors confirmed. He also gained a heightened appreciation of technology's potential, and of the macro-economic factors that would influence technological change.

8.4.4.4 Very significantly changed

Total 2 members: Group I, 1 member; Group II, 1 member.

Group I

S14 discovered that personal efforts alone were far more significant than he had realized, although much of the change came from effects on Heating of his planned move to a new apartment. S14 explained that he was amazed that just a few personal initiatives (e.g. moving to a new apartment and some other heating-related efforts) had an impact on his profile equivalent to future (trend) technological changes. Some of these technical changes are not trivial to effectuate, whereas his personal changes seemed comparatively trivial (especially since several were incidental to the move he was planning anyway).

Group II

S19 saw the greater impact and potential of personal efforts for his own household, which he (erroneously) extrapolated to private households in general.

Additionally, there were several other related surprises for subjects concerning aspects of their status quo profiles: Regarding flying, S7 was not surprised how much energy his flying utilized, but he was surprised how greatly it exceeded the Swiss average (400 km/person-yr). He thought regional or municipal flying averages (like Zurich's) would be much higher. (This issue is elaborated below in 8.8.2 "Energy communication: Subjects' mistrust of the data and model.")

S15 experienced a change in perception of how large her consumption was (compared to the average), if not in her perception of the factors driving it: She was shocked at how much more she consumed than the Swiss average in many categories. (She also thought she would be closer to her (statistically assigned) peer group whom she thought consumed similarly to her.) She was especially surprised because she had thought her behavior had already been "optimized" or rationalized. She noted that her basic needs like heating and food had energy values close to the Swiss average, while her leisure goods and activities like personal travel ("extras") were generally much higher than the Swiss average. Her distinction and differential finding between essentials and luxuries mirror Group II member S5's distinction between the same, which he termed "need to have" and "nice to have."

To summarize these Before/After questions and the preceding several sections, as a result of the interview session, subjects tended to rate external, less discretionary (especially technological) factors as somewhat stronger determinants of their energy consumption levels than personal factors. That is, after experience with the program, on average subjects had a slightly diminished view of the efficacy (or possibility) of personal interventions to reduce consumption. This was only generally true of total energy profiles. Individual sectors sometimes showed the contrary. Direct electricity consumption in living, diet, and public transportation (as well as direct fuel use for flying) were such counter-examples. For reducing electricity consumption from these activities, short to mid-term personal steps were on average more effective than broad-scale technological improvements like higher efficiencies, although this finding was true largely only of Group I (student) members. In addition, most subjects, especially Group II younger professionals, expressed a willingness to make modest efforts to conserve electricity for living and diet activity sectors.

Perhaps the most prominent counter-example was air travel, which was mostly recognized as highly important in one's energy profile.[87] Flying was seen to be strongly dependent on personal circumstances, behavior, and choices and only weakly influenced by technology. These two observations, the recognition of the prominence and importance of air travel in individual profiles and its dominance by personal determinants, held among most subjects along a range of self-rankings, incomes, and other demographic variables. However, as will be discussed shortly in section 8.4.6, recognition does not imply taking responsibility. The tendency to discount the meaningfulness of individual conservation efforts – and in fact the single consumer's (vanishingly small?) economic role in supporting and perpetuating the aviation market – is perfectly captured in the air traveler's standard defense: "The planes are flying anyway {whether or not I'm on board}."

Time$_1$/Time$_2$ and cross-group comparisons of views on the importance of personal vs. socio-technical changes for reducing energy use

On average, after experience with the program, subjects tended to rate external (technological or societal) factors as somewhat stronger determinants of their consumption levels than personal (behavioral and household investment) factors, with a few notable exceptions such as mobility.

The contribution of air travel

Experience with the program showed most subjects that long-distance leisure air travel is prominent (or even dominant) in their current individual energy profiles and that mostly personal-behavioral, rather than technological, factors drive any potential change in air transportation's future energy impact. These two observations, recognition of the prominence and importance of air travel, and its influence by personal decisions, were true across many subjects, demographic characteristics, and survey criteria.

[87] As noted, an extended CO_2 model would show flying several times more important on a CO_2 or climate-forcing basis than a pure energy use accounting.

8.4.5 Technological vs. social factors' effects on *national* energy use

In step 5 of the interview, the subjects set levels of the Type II social and demographic variables. Step 6 asked the subject to look at the combined effects and to judge between the two types, Type I and Type II:

"How do the two types of long-term variables, Type I and Type II, compare in importance? That is, which seem more significant to you for their impact on energy consumption?"

In short, the results from this question and from $time_1/time_2$ measurement show that after experience with the program, a strong majority of subjects identified Type II social and demographic variables as more important than technological variables for their impact on future national energy use. Scaled-up to the aggregate level, then, collective "behavioral"-type parameters again appear to dominate, while as we have seen, technological change appears on average more important at the level of the individual. As a learning process this reflected a moderate increase in social/demographic variables' importance (+13% total average) and a slight downward revision for technological variables (-6%). Subjects' judgment between the two sometimes clearly depended on their reference point, the goal they favored for future national energy levels (specifically whether reduction or stabilization).[88] The remainder of this section elaborates on these points.

Having chosen some value for the Type II parameters and observed their effects, subjects were asked to identify the most influential ones – in effect to perform an informal sensitivity analysis – and then to compare the impact of Type I and Type IIs. As this was a complex and challenging task, we subsequently simplified the comparison by offering to set all (or most) of the Type I parameters to their highest (most efficient) modeled levels while simultaneously setting all Type IIs to their highest activity levels (joint minimizations were also sometimes done).[89] The answers before and after this maximization exercise (if it was done) were recorded, and any change in opinion noted (reported below).

Here are some notable experiences and comments on this exercise:

Although the two classes are nearly balanced, S15 said, Type IIs are stronger and can cancel out gains from higher efficiencies. If instead we hold

[88] In addition, by shifting the focus from the individual to the aggregate level, this question implicitly brought out the distinction between efficiency and conservation: It highlighted the difference between smaller scale improvements in the efficiency of industrial production methods and macro-level increases in activity that defeat conservation aims.

[89] The interview question ran "We can make it easier to compare the effect of Type I efficiency variables to Type II social and demographic variables. We'll set both at high levels and look at the net effect on aggregate energy use."

down activity levels, we could have overall improvement in future energy use even with more moderate technological improvements than the maximum modeled. This assessment accorded with her earlier emphasis on behavioral changes over technological changes for the individual. Later S15 admitted the difficulty of changing social trends towards greater activity, especially in long-distance air travel.

S19 thought the maximization exercise lacked a grounding in reality. Efficiencies could indeed be maximized, he said, but Switzerland would not realistically see the highest (selectable parameter) levels of car use (American levels of driving (1600 km/person-year)), plane use (American flying levels (2200 km/person-year)), and other Type IIs. He suggested that better technologies were currently being adopted, but in Switzerland activities were not increasing nearly to the extent of the highest or even second-highest parameter settings. Thus, this subject seemed to suggest that Switzerland not aim to maximize efficiencies to compensate for Type IIs, since the latter will not reach the maximum levels simulated by the program. However, this conclusion assumes the national energy goal is stabilization, not a significant reduction from the status quo.

This illustrates how much people's conclusions depend on their points of reference. Two subjects could observe identical graphs, but while one said that Type I efficiencies were canceled out by Type II growth, the other might say that natural Type II growth was nicely balanced by improved Type I efficiencies. The first's aim or reference point is probably a future reduction in national energy consumption (progress towards 2000 watts per capita), while the second subject likely views stabilization at current levels (> 6000 watts per capita or 1400 petajoules nationally) as satisfactory.

An example of this in Group I was S18's observation that with both types of variables maximized, Switzerland's energy consumption in 2030 is just slightly higher than its current level. This reinforced his previous conclusion that Type Is were dominant. "We can have a luxury lifestyle and, with the best technologies, we can still keep our current levels of energy consumption as a nation. So, such good technologies are good for the environment." The subject seemed to view the 2030 trend (approximately 1600 petajoules vs. 1400 petajoules in 2000) as acceptable, so he justifiably concluded that without advanced technology, high activity levels would make things much worse.[90]

[90] According to his answer to a preliminary question on national goals, however, S18 favored a reduction in future national energy use to 70% of current levels. The judgment in favor of Type Is, then, might represent a change of opinion, misinterpretation, misrepresentation, or muddling of views on either this or the preliminary question. Or, he might have answered here on the basis of what he considered realistic to achieve or

In trying to decide between the two, S7 of group III eventually reversed his initial opinion and settled on Type II, experiencing something of a revelation in the process. Looking at the results of the joint maximization, he saw that the Type II variables completely overwhelmed the gains from improvements in Type Is, leaving total energy use roughly unchanged from current levels. The parameter "Percentage of one-person households" stood out for him as the most influential:

> *To have total energy improvement, we must have larger households {fewer one-person households} – more people living together. So it's the people, not the technology that's key for the nation! ... I'm happy I saw this ... I said in the beginning of the interview that it's only technology, but what I've seen here tonight shows me that one-person households {household size} make(s) much more of a difference than technology.*

Household size, he noted, was much more important than maximum possible (model-permitted) improvement in the other parameter he had singled out, auto efficiency: Moving from over 9 liters/100 km to 2-4 liters/100 km has less of a dampening influence on overall energy consumption than strong reductions in single-person households e.g. from 30% to 25% single-family households. He noted that the luxury of being able to live alone in an apartment was only possible in a wealthy country like Switzerland, and, secondly, he suspected that the total population and the percentage of single-person households were inversely related.

8.4.5.1 BA7: Comparison of strength of Type I vs. Type II influences on Swiss national energy consumption

This $time_1/time_2$ reading measures the change in a subject's assessment of the importance of Type I and Type II factors in influencing aggregate (national) Swiss energy consumption.[91] The total average of change for Type I was -6% (i.e. a slight downward revision in importance) and +13% for Type II (a moderate increase in importance of Type II factors). Group I

---

reasonable to expect, which could be higher than what he thought actually desirable. Despite a concerted effort to clarify the interview questions and answers through iterative refinement in preparation for the interviews, the ambiguities in this type of research methodology sometimes remained large.

[91] For these measures of change in assessments there is not necessarily a correlation between the subject's change in her assessment of Type I and her assessment of Type II: Unlike the judgment issue just discussed, the two types are not zero-sum, in that both can be, and can be considered, important.

yielded an average of -3% for Type I and +6% for Type II. Group II showed higher figures in the same direction: -11% and 14% averages, respectively. Group III lacked data except for one member (S7) who showed a 100% increase in his assessment of the importance of Type II variables, specifically household size (and a more moderate (in this case corresponding) decrease in his assessment of Type I's importance). These findings could be linked with subjects' assessments of the technical and the personal for their individual profiles, but that is not pursued here.

Technological vs. social factors' effects on national energy use

After experience with the program, a strong majority of subjects identified social and demographic forces as more important than the application of technology for the goal of reducing future national energy consumption. As a learning process, this reflects a moderate increase in the assessment of the importance of the former (+13% total average) and a slight downward revision in that of the latter (-6%).

8.4.6 Ranking key players' degree of responsibility for national energy conservation

One of the post-session "debriefing" questions (the final interview step number 11) was the following:

"In general, in the effort to reduce Swiss energy consumption, what measure of responsibility would you assign to the following actors? (Rank them from 1 (highest relative responsibility) to 7 (lowest responsibility).

Yourself
Your peers
Building owners and managers
Politicians
Technicians
Business and corporate leaders
Others

Feel free to comment or explain your choices, if you wish."

8.4.6.1 Overall summary of results

Average rankings:

Total: 1.7 Business/corporate leaders; 2 Politicians; 2.5 Building owners/ managers; 2.6 Technicians; 4 Yourself; 5 Your peers; 5 Others

Group I: 1.6 Business/corporate leaders; 2 Politicians; 2.5 Building owners/ managers; 3 Technicians; 4 Yourself; 5 Your peers; 5 Others

Group II: 1 Politicians; 1.6 Business/corporate leaders; 2 Technicians; 3 Building owners/managers; 3.1 Yourself; 4 Your peers

Group III: 2 Business/corporate leaders; 2 Politicians; 2 Building owners/ managers; 2 Technicians; 4 Yourself; 7 Your peers

8.4.6.2 Conclusions

There seems to be no meaningful difference between total results and group results, and thus no differentiation among groups on this question. The order of importance is business/corporate leaders, politicians (first cluster); building owners/managers, technicians (second cluster); and oneself, one's peers, and others (third cluster). Among some groups the order is slightly different but the general clusters are the same. Thus, decision makers at the highest level in business and government are held most responsible; then decision makers at the level of buildings and equipment, building owners/managers and technicians; and finally, with the least responsibility, the individual, i.e. the subject or her peers.

8.4.6.3 A note on question interpretation

This question extended the analysis beyond the main focus of the interview and was appropriately raised at the interview's conclusion. The question was in fact taken verbatim from the questionnaire for *Energy in Everyday Life*, the Swiss study described in the box at the start of this chapter's results section. To place it in the context of the present study, our interview concentrated on the effectiveness and loci of influences on energy consumption. One of the main purposes and best uses of the program as a whole is to allow the user to catalog and quantitatively to compare the effectiveness of interventions on either side of the user / producer-society divide. Much of the further discussion beyond this, in the interview and in suggested future applications, centers on the political feasibility or – especially for the aim of ecological modernization – on the mechanisms of implementation, of enacting those changes on the producer-society side shown to be most effective. Below (in 8.7) we report on subjects' thoughts on the political and cultural possibilities and limitations on the exercise of (their) influence to bring about such changes. Here in this question at the interview's end, we touched on the question of responsibility, which can be interpreted either as who can and should exercise their power to rectify the energy situation or who is responsible or "culpable" if they do not. Admittedly, the subjects could have interpreted the question either way or as a repetition of the earlier focus on effectiveness or ability; the question and results are not rigorously formulated. For example, it was clear soon after the interview with S14 (not in time for clarification before he replied) that he interpreted the question to mean what is in these actors' power to do, rather than what level of responsibility do they have; and other subjects probably interpreted some of both into the question. Yet all in all, if we apply *Energy in Everyday Life*'s maxim about conservation, "capability engenders responsibility", to the societal level, then perhaps we should not be so concerned about the possible conflation of the two.

The results here are thus not so surprising, especially if a number of subjects interpreted the question as one of effectiveness or ability in accordance with the previous interview questions. The results show an emphasis on upper-level policy makers first, then lower level decision makers, and lastly the individual. The top-down element dominates here as well.[92] As exceptions, some subjects (e.g. S17) ranked individuals first but only when viewed in the aggregate or in sufficient numbers to constitute a critical mass for change (viewing them this way eliminates the apparent scale inconsistency that troubled S21 (below)).

8.5 Hypothesis H: Separation of social from technological influences

Having implicitly suggested the independence of Type I and Type II factors from each other in a series of exercises and questions, the interview suggests a more subtle interaction (interview step 6):

> This program separated the two classes of long-term variables – technological on the one hand and social, cultural, demographic on the other – for technical reasons. But they could be treated together, and we just suggested some reasons for doing so {earlier in the questionnaire}. When examining influences on energy consumption, would you prefer a clear separation between the two classes of

[92] The results here match experiences with Dutch environmental focus groups in which participants "passed the buck," claiming it was the responsibility of politicians and business and not the individual, and least of all themselves [van Vliet, personal communication].

> variables or a combined treatment of both classes? For example, if you could design a program involving both, would you make them completely independent or, instead, dependent on each other?

In total, 12 subjects favored separation (i.e. keeping the program's presentation), six favored a combination or presenting both options, and two had no opinion. The program's separation of the variables biases answers towards separation, as it requires a stronger conviction to argue for changing to a combination. Still, we wanted to explore the issue of the interplay and co-evolution of the technological and the social, and this was the closest we could come within the limitations of the model and interview. This said, a number of interviewees had interesting insights.

No clear group-based pattern is obvious at this level of analysis, nor is it clear theoretically why there should be a difference between groups. Below, the results are discussed by group only for consistency of presentation and not for any particular analytic reason.

8.5.1 Group I

Five subjects favored combination or both approaches (jointly), eight favored separation, and one had no opinion. One subject in favor of combination (S9) agreed strongly with the suggestion of a close linkage between Type Is and Type IIs and said the combination should be like the joint Type I-Type II maximization exercise (see 8.4.5) but be done automatically. Other combination advocates agreed with the reasons suggested for combining but had no idea how it might be executed. Reasons the others gave for separating were that the two types are actually separate (S13); one cannot measure the effects of each if everything is combined (S8); the interplay is more interesting to examine when they are separated (S2); it is best to show separately two ways to approach the energy problem. Also, the two types are not directly causally related; rather, there are complicated connections. To show rebound, one can set the variables manually as in the joint maximization/minimization exercises (S21). Further arguments offered for separation were differences in how the two types can be influenced: Type Is can only be influenced by scientists and political actors. But Type II "lifestyle" variables are themselves only slightly amenable to political manipulation (S12 and S4); Combining the variables requires assuming a rate of dependence of Type II consumption levels on Type I variables, which is too difficult to predict accurately (S3).

8.5.2 Group II

Three subjects were clearly in favor of separation. One (S5) distinguished between purposes of scientific analysis, for which separation is appropriate, and presentation for a layperson, which may benefit from combination. As a layperson, the subject might have wanted to see a combination for the reasons suggested in the interview. But from an analytical or scientific perspective, he would not combine them since that would make isolating and interpreting the effects of any one factor too difficult.

8.5.3 Group III

One member of Group III favored separation while one admitted to being overtaxed by the question and lacking the knowledge or expertise to render a judgment on the matter.

Separation of social from technological influences on consumption

Subjects favored separation of technological from social variables by roughly a two-to-one margin. Subjects had varied and reasoned answers on both sides. Most of the reasons for separation were related to the function or operation of such an accounting software program and not generally because the subjects thought the two types were actually independent of each other.

8.6 Hypothesis G: Cross-temporal and cross-cultural comparisons

The following summarizes the interview questions for the thought experiments in interview step 7:

> Let's do a thought experiment. The lowest and highest values here {in the drop-down values for the Type II parameter "Auto kilometers"} represent, respectively, **a)** per capita kilometers driven per year in Switzerland in 1970 (3500 km) **b)** the distance currently driven per capita per year in the US (16000 km). Let's see the hypothetical impact on the Swiss average energy profile by imagining that Switzerland would **a)** return on average to a Swiss 1970s level of driving; **b)** emulate current US driving levels.
>
> (If you drive a car currently:) Imagine you lived in a Switzerland of the future in which people a) drove on average only as much as they

> did in 1970 b) drove as much as Americans do currently, reflecting a general shift in social norms towards, respectively, a) much less driving and b) much more driving.
>
> Would you go along with such shifts and change your driving levels accordingly, to the extent you could? To what extent?

Similar thought experiments were done for plane travel, simulating a) historically low levels (200 km per capita, actually only half the current national average) and b) current average levels in the US of 2200 km.

Like the questions concerning Type I-Type II separation or combination, the auto and plane thought experiments sought to use the capabilities of the program as a springboard to discuss issues otherwise difficult to quantify or illustrate. Douglas and Isherwood [1979] claimed that people synchronize their behaviors ("keep to the level") to conform with people whose behavior they see they see as relevant for themselves [Spaargaren 2000a]: Would our subjects change their behaviors to stay in line with a new (or old) "level"? But like the previous question, this one was only partly successful in execution. One problem, namely that national averages are not necessarily the proper reference groups for everyone, is discussed later (8.8.2). Another problem was that it was difficult for many subjects to imagine mass shifts in people's driving or flying habits in isolation, without being prompted by, or causing, reinforcing or counteracting shifts in other factors like prices or infrastructure. On this point they were correct, although for specificity and simplicity the exercise tried to examine the shift as a purely social phenomenon. Parallel with Brooks' experiences [Brooks 2001], many people do not readily think "sociologically" (or in this case "anthropologically" – see Chapter 2, section 3.3.2, "Anthropological/sociological treatments"): Subjects often reinterpreted the experimental change in independent variables to reflect anything other than our intended reversal of social norms, normalization, or cultivation (see Chapter 2, same section). Interviewees' assessments of the question (also reported on below) point to some of the difficulties.

Effects of changing social norms on individuals' behavior

Group I members would tend to follow a hypothetical escalating social trend towards more driving (although most because of a presumed concomitant decrease in public transportation services and the need to drive more to get around). Other groups are more resistant to change (or would even lower driving levels in counter-response).

8.6.1 Subjects' follow-up assessment of the exercise

In the follow-up questions at the end of the interview, subjects were asked to evaluate the usefulness of these thought experiments involving air travel and driving: "Did the thought experiments with historically low Swiss auto travel or high American levels of driving and flying help shed light on alternative development patterns?" [i.e. explore another "set of socio-technical alternatives" (Chapter 2, section 4.2)].

Viewing subjects' responses as a whole, seven subjects answered clearly in the affirmative. Ten subjects answered unequivocally no. Several of these latter pointed out, justifiably, that the question was not examined in any great depth in the exercises. A number repeated their criticisms that North American (or past historical) scenarios were not realistic development patterns or models for Switzerland, and others again insisted that the notion was too hypothetical and that it was too difficult to know how one would behave under such imaginary circumstances.

Some five subjects hedged their answers. S17, for example, attributed the difference in driving levels between Switzerland and the US to the difference in land prices, which is a matter of supply, and to different zoning laws, which are also largely a function of geographic conditions.

8.7 Consumer-citizen involvement in affecting less-discretionary forces

8.7.1 Hypothesis F (completion): Elusiveness of the 2000 watt per capita society: Why is future Swiss national energy consumption so large (even with optimistic assumptions)? Is this inevitable?

In (optional) interview step 10, one or more of the subject's personal energy profiles (i.e. status quo or short- or mid-term conservation profiles) were scaled up to national levels, under current or various possible future technological conditions. Discussion of macroeconomic scale is not often pursued in the study of sustainable consumption (and is unheard of in policy settings). This was an opportune time to explore the issue, especially via a function that let users gauge their contribution to the aggregate if their choices were to be widely imitated (assuming no social trap/prisoners' dilemma problems). It also allowed us to explore the other side of Hypothesis F, "Aggregate data comparisons," by making explicit links between the individual and the aggregate levels and helping visualize a flow this time from the former to the latter. In general, it was indeed "useful and instructive" to make these links: The scaling-up function generated much thought and discussion (and even, as anticipated, some degree of dismay and resignation) from those subjects who explored it.

The subjects' resulting scaled-up graph was invariably at least the equivalent of 4000 watts per person (for the 2030 projected Swiss population), even for the most spartan or conservationist-minded Group I student, and even at the highest modelled levels of technical efficiency.[93] Finally, subjects were asked: *Why is national energy use still so large? Why aren't combined behavioral, social, and technological changes enough even to approach the designated ecological standard of 2000 watts?* (Which categories are still so much larger than they used to be in the 1950s when Switzerland consumed 2000 watts per capita?)

This topic stimulated an outpouring of rich "lay social theories of a local world – the frank reflections of ordinary {sic} people forced to think through that world" [Lemert 1999]. Certain parallels to scholarly social science's treatment of consumption (Chapter 2) are apparent. The box below abstracts from the themes in subjects' responses.

The elusive 2000 Watt Society:
Why is future Swiss national energy consumption so large even with optimistic assumptions about technology and activity levels?

Themes in the subjects' discussion: (Partial) disbelief in the feasibility of the threshold; techno-socio-political barriers; the need to reduce grey energies; perverse economic incentives and the huge transport of constituent materials as inflators of grey energies; lifestyles of luxury, availability, and convenience; lifestyle normalization at ever-higher levels and the (im)possibility of reversal; far-flung leisure travel; shrinking household size; low public awareness; wants emphasized instead of needs; probable need for draconian measures; and technological fixes possible through changes in the technological paradigm.

[93] The Type I parameters need much higher modeled value(s) at the top end and/or the Type II much lower value(s) (including of course population); otherwise Switzerland cannot approach 2000 watts per capita. The program's modeling shows Ehrlich's IPAT identity at work. We did not dwell on the 2000 watt target or heavily emphasize it until this optional part at the end; doing so might have demoralized the subjects and distracted them from the other exercises.

8.7.2 Shaping future Swiss consumption trajectories

> [The following question asks for a community perspective.] Do you agree (1) that Swiss society has the power to make a collective choice of its consumption levels? Or rather do you agree (2) that social and economic processes have their own internal mechanisms that will automatically or inevitably lead to increases in energy consumption?
>
> [The following question asks for a corresponding individual perspective.] Do you feel you have some personal ability to influence – "decision making capability" over – your energy consumption?

These first two interview follow-up questions (interview step 11) explicitly state that the questions call for a community and individual perspective, respectively. The first one is essentially a question of whether the national trend levels for Type I and Type II variables are inevitable or not, i.e., in terms of the program function, whether the society can craft a "Swiss 2030 Tailor-Made Scenario" different from the "Swiss 2030 Trend" graph (the third and second graphs, respectively, in the lower left portion of Figure 4-1). The previous section 8.7.1 showed that subjects recognize the difficulty of radical reductions. Here we ask effectively whether even a moderate deviation from projected increasing trends is feasible. These questions were also adapted from similar ones in *Energy in Everyday Life* and *Democratizing Environmental Policy*.

The overall group results show that a strong majority agrees with (1), that Switzerland has the power of choice. There were only two apparently unqualified endorsements of (2). However, many who answered (1) emphasized that the society may have the power, but it will not necessarily exercise it. Overall, the interviewee pool seemed to have a greater sense of social "*Machbarkeit*" (flexible feasibility) and less of a feeling of helplessness in the face of corporate and social trends and policies than the majority of focus group subjects had in *Democratizing Environmental Policy* or interviewees in *Energy in Everyday Life* (although it is prudent to recall here that our subject pool, and probably the ones from *Energy in Everyday Life*'s qualitative interviews, were not necessarily statistically representative of the population).

Shaping future Swiss consumption trajectories:

A majority of subjects felt that Swiss society has the ability to make a collective choice of its energy consumption levels and that trend level increases are not the inevitable result of "automatic" socio-economic processes. This does not imply they think that ability will be exercised.

8.7.3 Affecting "less-discretionary" forces and the role and influence of the consumer-citizen

A summary of subjects' views on the issue of the tractability of Swiss energy consumption, and the extent of the end-users' potential involvement or role in shaping it, is provided by the answers to this follow-up question:

"What do you think your role is in addressing {societal} problems connected to energy consumption?"

Viewing all groups together, 12 subjects answered unequivocally that they had no role. Two or three others found it hard to say definitively. Among the latter, S17 said it may be possible to influence others, but not effectively as an individual (voter). Social changes do not come quickly, and creating a constituency for change also takes time. The process is aided by people's seeing and feeling the adverse effects of excessive consumption in the form of noise and air pollution, for example.

The remainder believed they had, or might have in the future, some individual influence, but mostly not as individual consumers. Several students of environmental engineering and architecture counted themselves in this group because of their future professional roles. Decision makers like politicians, policy makers, key researchers, and corporate officers and business managers could exercise greater influence and therefore had larger potential roles, according to the majority opinion.

A few subjects specifically invoked their roles as consumers or citizens participating in or supporting the activities of NGOs.

In addition, interview step 9 gave subjects an opportunity to brainstorm on the possibilities for greening infrastructures, technologies, and social practices connected to a sector of their choice. In what way and how easily could the relevant Type I or Type II variables be changed to favorable levels; and (to expand on the theme of the previous section) what are end-users' and upper-level decision makers' intervention points? We omit the details and summarize the general findings in the box below.

Consumer-citizen involvement in addressing Swiss energy consumption:

Twelve subjects said unequivocally they had no means of personally helping influence the national energy situation. Three couldn't answer. The remainder believed they might have some influence in the future, but not generally as individual consumers. Several students of environmental engineering or architecture viewed themselves as potential future decision makers in their career capacities. A few subjects specifically invoked their roles as consumers or citizens who participate in or support the activities of NGOs.

8.8 Energy Communication

8.8.1 The program and interview as a communicative and educational tool: Subjects' degree of learning over the course of the session

Communicating about energy use and facilitating learning, broadly construed, were main objectives of the modeling and interview effort. Assessments of subjects' learning have been described extensively above in reports on (explicit or implicit) Before/After ($time_1$/$time_2$) questions. This section describes the outcome of the few remaining Before/After tests and presents other material pertaining to energy and risk communication.

8.8.1.1 BA1: Subjects' self-rankings as energy users (high, medium, low) vis-à-vis the Swiss average and lifestyle group averages

Subjects' $time_1$/$time_2$ self-assessment of energy consumption levels compared to both the Swiss average and the average for their respective lifestyle groups:

Group I learned on average that their energy use compared to their lifestyle group's was slightly higher than they had thought. Group II learned on average that their energy consumption was considerably higher than they had thought, measured against both the national average and that of their lifestyle group. Group III showed no change in self-assessment of their energy consumption, i.e. good accuracy from the beginning.

8.8.1.2 BA3: Subjects' assessments of the environmental friendliness of their lifestyle as measured against an ecological standard (2000 watts/person)

The total average is about -1.11, i.e. a downward revision of more than one assessment "unit" (e.g. from highly friendly to moderately friendly or from moderately unfriendly to highly unfriendly). All three groups have a close average of about –1; all have approximately a one unit downward revision. Group I's average is -1.21, the largest in magnitude, suggesting that students overestimated the friendliness of their lifestyles the most and/or learned the most about this fact from the session. Group II's average, -1, is the next largest in magnitude, but two of its four members had no change. Group III's average of -.75 comes from only two members (one could not answer the question): 0 and -1.5.

To contextualize, the gap between people's perceptions of their environmental performance (their environmental image of themselves) and reality (the impacts of their practices) is regularly observed, as is the gap between people's avowals of environmentalism and their actual behavior or the environmental impacts of their lifestyles. Drawing subjects' attention to these gaps (at least as far as energy use reflects environmental performance) in the hope of reconciling or narrowing them is a mainstay of Energy- and Environment-Revealing approaches. We are naturally more skeptical that such information provision will actually lead to bridging the gap; however, as part of a joint Energy- and Social-Revealing program, it may be a first step in that direction.

8.8.1.3 Correlation

BA3 and **Personal Income:** 0.40. This barely moderate correlation suggests that higher income earners experienced slightly greater learning through the program concerning the (un)friendliness of their lifestyle, i.e. the difference of their energy balances from 2000 watts. This is not surprising, as the higher income earners tended to have larger status quo profiles, so the divergence, and evidently the visible starkness of this divergence, was larger among these subjects.

Time$_1$/time$_2$ self-assessment of the environmental friendliness of subjects' lifestyles

On average, subjects learned that they had somewhat overestimated the environmental friendliness of their lifestyles.

8.8.2 Energy communication: Subjects' mistrust of the data and model

The believability of the data and/or the embedded model for the end-user is an aspect of bi-directional communication that surfaced a number of times during the interviews.

S19 was amazed and perturbed that his total energy use so greatly exceeded the Swiss average. To reassure himself, he tried to cast doubt on the accuracy of the statistics.

Many subjects were surprised at how low the Swiss average was for air travel (400 km/year). (As with all of the national averages, this includes the entire population, young and old alike.) S6, however, thought this figure was impossibly low and refused to accept it. He preferred to see the flying statistic for his particular lifestyle group instead, which he understandably thought he could better relate to his experience. However, the program did not make this datum available. In retrospect, this omission was a clear deficiency of the program and the session. Sociologists know that most people tend to relate their behaviors (and any changes in them) better to those of their peers, not (only) to an abstract national average. This is clearly an area to enhance in future versions.

Several others, S7 and S19 among them, also challenged the national monthly average for miscellaneous expenditures (40 Swiss francs) as unreasonably low. Such data, they said, should be differentiated by region or earning bracket: They knew no one who spent so little per month on miscellaneous costs in Zurich (where a single visit to the barber can easily cost as much or more).

Several subjects justifiably raised the issue of the believability of the model on which the program was based. S5 qualified his answer to the general question of how much he had learned from the session by saying, in effect: 'If the model is true – if I can trust what it has shown me to be a fair representation of present reality and future scenarios – then I have learned such and such.' For such end-users – as opposed to the modelers – the model is an impenetrable black box that they must take on faith in order to have confidence in the conclusions they draw from it.

Indeed, the subject pool's other economist, S17, cast doubt on the model's believability in some respects. Confronted with a future tailor-made Swiss national scenario for energy use that far exceeded the 2000 watt threshold, even with optimistic technological parameter settings, he countered that the model presents only a few possible future realities based on values diverging (but not greatly) from extrapolated trend values, and while these trajectories may be likely, they are not the only possible future scenarios.

Paraphrasing our exchange, the subject said that development could go in another direction, that we could witness technological change that allows people to do far more with less. I pointed out that these improvements were already modeled and projected in the program. He retorted that those were just trend extrapolations and did not capture the more fundamentally and qualitatively different technologies or technological systems that might come into use. These would make much more difference to the national energy account. 'Fundamental shifts {changes in techno-economic paradigms} have taken place in the last one hundred years ... The program's modeled scenario is one possibility, but there are other scenarios. In general, all of these sorts of long-run forecasts have proved completely wrong in the past.' (S5 and S17 are right, of course: There are different modeling approaches with potentially different results and implications. For an example of one such model, from the field of artificial societies, see footnote 96 in Chapter 5.)

In essence, then, S17 cut the Gordian knot of energy consumption with a strong technological fix reminiscent of eco-efficiency and other contemporary lines of environmental management thinking (although he also suggested the possibility of behavioral shifts prompted by higher prices reflecting growing environmental concerns). S17 divined that environmental sustainability would require either much greater technological improvements than the best the program offered, much less activity, or both (more accurately, socio-technical shifts that brought about deep changes affecting both); not surprisingly, given his educational background and his experience in the session, he chose the "technical" route.

Users' (mis)trust in the data and model

The believability of the data and/or the model, and users' trust in them and faith in the conclusions they draw based on their experiences in the session, were important issues for some participants. Several challenged data on flights or monthly expenditure (these seemed extremely, even unreasonably, low compared to the subjects' own). A number wanted greater opportunities to compare themselves with regional and/or lifestyle group data. Some did not believe energy from flying would be, on net, largely unaffected by future technological change. Others qualified their conclusions in the session by saying "if the model here can be trusted." One subject in particular, confronted with future tailor-made Swiss national scenarios for energy use that far exceeded the 2000 watt target threshold, even with optimistic technological parameter settings, insisted that the most advanced technological changes described by the model represented only one possibility based on extrapolations and that "quantum leap" improvements over these could well be expected.

Several subjects, including S18 and S6, challenged the finding that the energy consumed per kilometer by air travel would not noticeably improve in the future from any possible manipulation of Type I technological parameters offered by the program. Although perhaps surprising, this result was verified by the modelers and attributed to the reduction of future plane size and occupancy rates, which will negate the savings from any improvement in the efficiency of jet engines [Dürrenberger, personal communication]. Such explanations did not satisfy S18 and S6, whose insistence that the unchanged flight-related energy consumption was incorrect amounted to a challenge to an aspect of the model.

8.8.3 Other energy communication issues: Misleading, misplaced emphasis in the presentation of technological and social factors

In reviewing the logic and effectiveness of the interview structure and analyzing some related reactions from interviewees, we discovered a danger that the first part of the non-discretionary section (interview step 4) devoted too much time to Type I technological factors in isolation and only showed their effects on personal profiles. Despite the subsequent attempts at a balanced presentation of the effects of Type II variables (on the average and aggregate), some people seemed to remember only the preliminary lessons about the influence of technology and to "take home" only this message. It was not our intention to cultivate an exaggerated belief in technological fixes.

In S7's case, the order, repetition, and slow build-up through the Type I section seem to have had the positive pedagogical effect of increasing the force of the surprise in the section on Type IIs (see the discussion of S7's "revelation" in Group III of section 8.4.5 above). By interview step 6, where he saw the action of Type IIs on the national level, S7 had reiterated his technological optimist's answers several times since the preliminary questions in step 1, making the change in his own opinion at this point seem all the more stark to him – it accentuated the final lesson more by bringing his previously hardened technologist opinion, reinforced by the early experience with the program (with his profile only), into greater relief with his changed opinion later on.

Generally, the danger of overemphasis may be overcome by the proper balance, order, and equal presentation of Type I and Type II variables. The shortened and expedited presentation in the streamlined interview versions with businesspeople probably better served this end.

8.9 Subjects' evaluation of the program and the interview session

8.9.1 How useful and instructive was the session?

Here we summarize the results of follow-up questions evaluating the usefulness and educational value of (1) the program as a whole, (2) exercises involving the generation and analysis of the subjects' conservation profiles, and (3) the sections involving Type I and Type II influences on energy consumption.

The **total** combined average was 94% (overwhelming) approval of the program on these points.

8.9.2 Did the session bring out various points?

Individual summaries were compiled for these questions:

"Do you feel that energy accounting programs like this one help clarify:

1. How your personal behavior affects your energy consumption?
2. How your lifestyle (affiliation) affects your energy consumption?
3. The variety of other factors that affect your energy consumption?
4. The relative importance of these factors?
5. The connection between your consumption and Switzerland's as a whole?"

Summaries and averages of these (numerically converted) yes/no questions yielded the following:

Total average: 87.5% Yes

Subjects' evaluation of the program and the interview session

A large majority of subjects felt the program was highly instructive and useful for themselves.

8.9.3 How complicated did subjects find the program and graphical user interface?

All subjects claimed the program was understandable. Most (15 subjects) ranked it "not complicated" while some among them noted that certain interview questions built around the program were complicated. At least four subjects ranked the program moderately complicated. One (S15) ranked it moderately-to-highly complicated. The interview built around the program, however, added further layers of difficulty. Some subjects admitted this openly, while others' inconsistent answers on certain themes pointed to the challenging overall nature of the interview session.

8.9.4 Usefulness of the program for informing and educating the general public

"As far as information and education efforts {for laypeople} go, do you think this program is a helpful approach to addressing energy consumption?"

The majority answered with at least a qualified "yes." Group I offered mixed answers. Group II generally said "yes," especially for (upper-level) students in a suitable educational environment. Group III members had some reservations about endorsing it for general use.

Viewed generally, five or six subjects gave at least partial "no" answers. Several of these said that the program section with Type I and II less discretionary variables was too difficult and required specialized knowledge (a comment on content). Others wondered how many laypeople would be able and willing to grapple with the proffered form of graphical displays (comments on the communication medium). Some subjects gave unqualified glowing endorsements. Some said that certain parts were easier to follow while other parts would have to be simplified or shortened for general use. Some endorsed it for older students (e.g. at a high school level), but for younger (or less advanced) students, other tools would have to be devised.

Usefulness of the program for laypeople

Most subjects felt the program would be a useful tool for informational and educational purposes with the general public, perhaps given certain changes or simplifications, or with advanced high school students and/or in a suitable educational environment. Some felt the section with technological and social influences was too complicated for such a purpose, and that therefore only the personal, purely household-oriented module was suitable for use with the general public.

Of course, if the module with the long-term technological and social parameters is jettisoned, we are back to where we started, in the company of plenty of purely household-based energy accounting software currently available for the public. Somewhat humbled, this is an apt point to conclude the report on the experimental results and turn to broader conclusions.

Chapter 5

ACHIEVEMENTS, OPEN QUESTIONS, AND LESSONS LEARNED

This concluding chapter takes a step back from detailed empirical reporting and selectively reviews what has been learned in this study and how it might be used for sustainable consumption research and applications. To some extent it crystallizes findings from the empirical chapter 4 in light of frameworks and concepts from the more theory-oriented chapters 1, 2 and 3. This chapter does not attempt a systematic summary of results. Rather, it highlights achievements, revisits prominent themes in light of the fieldwork, makes recommendations for policy and research, and poses questions for consumption theory and practice.

1. WHICH QUESTIONS HAVE BEEN ANSWERED?

In the experimental section of this study we set out to examine the viability and usefulness of a certain energy communication approach for the public.[94] The results are wide-ranging and could be used to support a variety of conclusions and to examine a variety of additional premises. The leitmotifs that sounded variously in the foreground or background

[94] Empirical findings were also examined in the review of past Swiss studies detailed in Appendix A in [Goldblatt 2002] and briefly described in the box at the beginning of Ch. 4.

throughout these chapters – the individual's relationship to the collective, actor vs. structure, risk communication to spur socio-technical innovation, public deliberation and participation in environmental policy, and others – are the subjects of many more thorough studies than this one. The major innovation here is the novel combination of such themes in a trans-disciplinary application for sustainable consumption, one that brings a dual individual and institutional-structural perspective to the end-user in a quantitative and holistic fashion.

1.1 Top-down or bottom-up?

At the end of the search for effective intervention points for consumption in Chapter 2, the questions were posed "Which of the two broad types of incentives to alter {the present trajectory of unsustainable} consumption, internal 'bottom-up' or external 'top down,' are the most motivating and most durable" and, relatedly, can individual behavioral change alone work, or is institutional and social change necessary? The favored sociological construct, based on Gidden's structuration theory, in effect answers at least the second one *a priori*: A structural impetus is also necessary.[95]

The new configuration of the ECO_2 energy accounting model discussed in this book, the ECO_2–interview version, brought technological and social systems into a dynamic interplay with householders' individual behaviors. It let users discover for themselves how nuanced and changeable the threshold is between discretionary and non-discretionary, in different individual life situations and at different levels of aggregation. Concretely, subjects might observe that their diet, for instance, consumed more (combined direct and embodied) energy than their car, decide for themselves whether they could make any changes in their eating habits, and then discover perhaps that improvements in industrial processes or the power generation mix had a greater dampening effect than any dietary changes they might reasonably consider. In the aggregate, however, they might then find that Swiss trends towards increasing auto travel, and at low vehicle occupancy, brought collective automobile-related behaviors and norms back to the fore and Swiss energy expenditure on private mobility back to a dominant place in a national energy accounting.

A small majority of subjects from the data sets examined here (as well as those in the two past Swiss studies) seemed to put somewhat more emphasis on top-down changes overall; however, by the interview's end, subjects in

[95] As noted, this means that the introductory framing hypothesis A: "Freedom of Choice" was more or less confirmed *a priori* according to the sociological outlook adopted at the outset.

the ECO_2 sessions could not in good faith entirely "pass the buck" in terms of individual effectiveness and responsibility. This is expanded on in section 3 below.[96]

1.2 Metric for gauging experimental success

As discussed in the annotations to Hypothesis E in Chapter 4, we grappled from the start with the question of how to measure success in the experiments with the software and interviews. Clearly, we could not necessarily expect a commitment from the subjects to make behavioral changes (only a few expressed a new resolve to make changes in the home, and that resolve points to an enhanced behavior *intention* that may not be translated into practice anyway). Chapter 3 took pains to explore the theory that elaborates the conditions – personal, situational, and structural – that are necessary for the translation of enhanced knowledge into behavior. A few participants in the ECO_2 interview sessions specifically cautioned the interviewer that their new-found understanding of the factors and issues surrounding energy consumption would not lead them to make any changes in their personal consumption choices or habits. Nor of course is that the only positive response that could be hoped for. The various forms of end-use

[96] Work in the new field of artificial societies (e.g. Gilbert Nigel and Rosaria Conte (eds.) (1995), *Artificial Societies: The Computer Simulation of Social Life,* London: UCL Press; Rauch Jonathan (April 2002), "Seeing around Corners", *The Atlantic Monthly*, vol. 289, no. 4) is shedding light on the divergence between properties at the micro and the macro levels. Findings from its unconventional sort of modeling may challenge aspects of our information approach for sustainable consumption on both a theoretical and empirical level. In A-society computer simulations, new and often unexpected properties and discontinuous shifts "emerge" on the system level when individual agents are modeled to interact with their neighbors according to simple sets of rules. Positive social transformations often depend on modeling individuals as both diverse and limited in their knowledge of aspects of their immediate surroundings, i.e. qualities of actors in their immediate vicinity. This seems to run counter to the aim in the research presented in this book to increase people's knowledge of the whole system (and the conventional environmental dictum that one should "think globally and act locally"). We have noted the merit in interview subjects' requests for more data on their peer groups and fewer perhaps about "society at large." Of course, our provision of information on the national level, in terms of aggregate and average statistics, is a far cry from making people "fully informed." And the complicated cognitive, psychological, social, or political uses to which people might put this knowledge are beyond the range of these simple modeled worlds. However, artificial society work also suggests that certain small, targeted interventions can have large, discontinuous results on the macro-level, which argues perhaps for a consumption approach more oriented toward "key" players than toward the masses. The two approaches co-exist well and in fact should be jointly pursued.

involvement for the greening of consumption institutions are even more diffuse in space and time and would be harder to measure, although one could imagine – especially in a future application pairing the use of the enhanced ECO_2 with focus group discussions of specific involvement issues – a post-session questionnaire that asked whether the participants would now be willing to lend support to green consumption: Household conservation in sector x, green taxes, purchase of green electricity, environmentally-oriented referenda, and so on. This was pursued to some extent in Dahinden's (2000) *Democratizing Environmental Policy*. Discussions are underway with Swiss energy agencies to develop sophisticated, full-scale versions of ECO_2 that draw on large municipal-level databases and can simulate the effects of highly specific political measures [Gregor Dürrenberger, personal communication].

1.3 Success in researching and applying domestic sustainable consumption

The consumer-oriented Eco-team meetings used by environmental groups in many countries are based on an Environment-Revealing monitoring effort that allows people to track their resource consumption and waste generation, and compare themselves and exchange conservation tips with members of peer groups. According to the program's founder Gershon, people ask four basic questions about how to change their behavior to help the environment: 1. Where do I start? 2. Which are the most important actions? 3. How do I do them? 4. Will it make a difference? [van den Burg 2001]. By combining Energy-Revealing and Social-Revealing approaches, the ECO_2 sessions provided participants with opportunities to examine, and at least partially answer, questions 1, 4, and especially 2. For this last one, the program allowed users quantitatively to compare the effects of different actions on their energy profiles. Using the aggregate displays and the function that scaled the user's profile up to national levels, they could see what a difference various steps would make and confront issues of critical mass and social traps (question 4). As emphasized, question 3 largely awaits further studies, social-psychological internalization, and political actualization of the lessons participants took home from their experiences during the session.

The empirical research presented in Chapter 4 also covered three of the four basic agenda items proposed by Spaargaren and van Vliet for consumption-oriented research into environmental innovations in household contexts: Lifestyles; evolving or escalating standards of comfort, cleanliness, and convenience; modes of provision of public and private goods and services as well as production; and domestic time-space structures

[Spaargaren 2000b]. The last one, typically examined in studies of the sociology of technology, was only minimally achieved and is perhaps best reserved for practitioners in that field.

The success in expanding the set of determinants in an end-user consumption information tool beyond the usual discretionary household variables like lighting and thermostat control should encourage energy analysts of the Social-Revealing school. Extensions of the software and interviews could be used to stimulate discussion in both public and policy circles of problematic trends for energy consumption like technological lock-in and perpetual demand escalation in private transportation, for example (See "Suggestions for further research" below). According to SCOT theorists, this may be the start of the process by which the perceived "non-discretionary" becomes open to the possibility of collective change [Goldblatt 2003].

The inclusion of embodied energies was another innovative feature of the software and the interviews. The interview sessions showed participants how large a proportion of the current and future overshoot of Switzerland's designated sustainability threshold of 2000 watts per capita could be attributed to grey energies, even with considerable technological progress and conservation-favoring social changes. Including embodied energies helped meld the program's presentation and interface to the larger consumption perspective and social science framework that lay behind the interview sessions [Goldblatt 2003].

Our approach also answers the United Nations Environment Programme's (UNEP) recent call for strategies for sustainable consumption that extend "beyond the conventional economic considerations, to include individuals as *citizen*s, not just mere *consumer*s" [Manoochehri 2001].

A recent UNEP report entitled *Consumption Opportunities* shows a striking propinquity with the theories independently advanced here. *Consumption Opportunities* places consumption "optimization" on equal footing with dematerialization as a vital strategic element for sustainable consumption. Optimization consists of consumption that is *different, conscious, and appropriate*, where *different* entails changing the institutional infrastructure to allow for different consumer choices, *conscious* involves greater individual consciousness in "choosing and using," and *appropriate* addresses drivers and levels of consumption.

Furthermore, their conception of "systemic" sustainable consumption seems remarkably close to the particular communications effort espoused in this book:

> Sustainable consumption in this 'systemic' sense … engages, economically and socially, from the bottom up, using the actions and perspective of consumers and citizens as its starting point, rather than the big-picture assessments of the global environment of sustainable development discourse … Indeed, one of the shortcomings of sustainable development since the Rio Earth Summit has been the difficulty in practice of bringing down to the level of communities the visions and commitments elaborated in intergovernmental circles. A 'systemic' approach to sustainable consumption is likely to overcome this, by starting at the personal and branching out to the broader issues.

Naturally, UNEP still holds government and industry primarily responsible for structural changes that promote "different" consumption and describes as "misleading" suggestions that consumers could drive this process. However, they suggest consumers, communities, and civil society as the major agents for "conscious" and "appropriate" consumption, which is very much in line with this book's promotion of the potential for end-user-oriented monitoring for environmental reform.

1.4 Pedagogical use and timing

Overall, the results of the pilot study offer an endorsement of a broadened information approach for sustainable consumption. The Social-Revealing approach, at least as expressed with a tool like the ECO_2-interview version, shows some promise for inducing, if only indirectly, personal conservation, consumer-citizen involvement, and (re)consideration of social practices, with the following qualifications:

Realistically, on close inspection in a follow-up session, many interviewees would likely admit that however interesting, informative, or entertaining they might have found the sessions, the lessons they learned have seen no lasting practical applications in their lives. This drives home the necessity of reaching people earlier in their social and cognitive development so as to increase the chances they internalize and later actualize what they learn. However, the complexity of the picture presented sets a lower bound on the age at which all but the most precocious could be expected to follow or relate to the software's presentation. It is not clear whether it matters that the crucial window for influencing the moral development of a child may normally close before the child is old enough to grapple intellectually with the issues we are concerned with here. This is a question for social psychologists, educators, and ethicists.

Thus, a program like the ECO_2 might be introduced into mid- to late high school environmental curricula, perhaps earlier, depending on the capabilities of the students. As one enthusiastic interview participant put it: "This {program session} is something everyone should experience at least once in their lives." If so, it should ideally be during adolescents' formative years, at the latest, so as to yield the most in behavioral, societal, and political dividends.

For pedagogical use with older, established householders, one might look for times when domestic routines have become temporarily "de-routinized," for example following utility disruptions, energy crises, strikes, unusual weather events,[97] and the like, when the experience of using the program might actually contribute to a change in direct consumption behavior in the household. These episodes may increase the chance of greater reflexivity, on a personal or social level, according to theory at any rate: Van Vliet and Spaargaren's events that de-routinize ordinarily routine household behavior [Spaargaren 2000b]; moments for collective historical reflection and alternative choices [Wilhite 2000]; and Bijker's (1995) reversing technological closure; Wilk's (1999) cultivation process, corresponding to some unusual event, "of opening existing needs {inscribed in the habitus} to question, discussion, and debate." In any such event, prior experience with a tool like ECO_2 may also show the individual to have been primed to participate in a political process of change.[98]

[97] Especially the increasingly frequent deranged weather phenomena like record-setting temperatures, storms, floods, and droughts that suggest accelerating climate change.

[98] During the 1991 Persian Gulf War [see Goldblatt 1993] and in the aftermath of the devastating terror attacks on the US on September 11 2001, the American nation experienced crises that could have led to reflexivity on energy use, especially since sizeable portions of petrodollar earnings in the Persian Gulf were and continue to be funneled into the financing of terrorist groups and regimes there. National security issues could have opened up the discussion to consideration of the bloated American demand for energy, even without reference to the associated environmental burden. Alas, political tunnel vision (and associated government corruption) has emphasized the usual, tired supply-side approach. This has only reinforced the cultural sense of resource entitlement. In addition, it appears the Clean Air Act and other environmental legislation have been rolled back under the dissimulating veil of increasing indigenous energy production (especially coal) as the EPA has lost its tussle with a corporate-dominated Energy Department (see C. Drew and R. A. Oppel Jr., "How Industry Won the Battle of Pollution Control at E.P.A.," *The New York Times*, March 6 2004.) Switzerland, with its relatively higher national environmental consciousness, seems to be in a better position for the sort of reflexivity needed for sustainable consumption, but here too the political winds and zeitgeist blow foul.

2. LIFESTYLE GROUPS AND DIFFERENTIATED APPROACHES TO SUSTAINABLE CONSUMPTION

Chapter 4 described how interviewees were grouped and how questions and themes were analyzed with respect to these groups and to the whole. These groups corresponded roughly to (one of many possible sets of) lifestyle groups whose members could make differential use of options "in the ecological modernization of sectors 'in a way that fits their lifestyle'" [Spaargaren 2003]. Groups I, II, and III showed themselves to be clearly differentiated in their use of energy-consuming devices and services, in their direct reduction potential (willingness to change), in their recognition of the role of factors higher up on production-consumption chains and networks, and in their possible willingness to participate in the greening of these chains and systems.

This implies that different lifestyle groups might use or support certain consumption interventions or "reduction strategies" preferentially to other strategies, although preferences here may not consistently correlate with other fixed lifestyle group characteristics. Still, assessing householders with such a tool as the enhanced ECO_2 can reveal which broadly defined lifestyle group they belong to and thus, for policy purposes, which clusters of strategies might appeal to them; and, in applied research, which strategies might be practical to develop with them in a further interview or focus group setting. Van den Burg is pursuing this avenue for Dutch utility consumers (see section 5 below).

3. PERSONAL RESPONSIBILITY

Considering the first few research hypotheses on people's differentiation between controllable and non-controllable influences on their consumption, the two past Swiss studies [Dahinden 2000], [Bovay 1987] reviewed reported some cases of people who believed they had no discretionary power over their energy consumption at all. Some of the ECO_2 session subjects, especially from Group III, held a similar opinion, at least at the start of the interview, in that they claimed they could not change anything about their activities to lower their energy use. However, few to none claimed that this was universally or collectively the case, since most agreed both that Swiss consumption was not purely technologically determined and that across a long enough time horizon, Switzerland could chart its own energy future.

The study described in *Energy in Everyday Life* emphasized the issue of personal responsibility in its interview questionnaire. The ECO_2 interview focused more on differentiating conditions where either personal or socio-technical interventions were more effective, and only in the interview follow-up questions did it explicitly raise the issue of responsibility.

Farhar and Finger's (1994) studies suggest different levels of recognition, understanding, and reaction to less discretionary constraints among the American and Swiss populations. The resignation, fear, and helplessness in the face of external forces Finger reports among the Swiss seem characteristic of a portion of the subjects in the two prior Swiss studies, but somewhat less true of the subjects generally encountered in Chapter 4, fortunate for our hopes for expanded citizen-consumer involvement. In designing the program and interview sessions, did we manage to avoid the danger of encouraging people to "pass the buck" when it comes to personal responsibility for conservation or restraint? In several places we noted the possibility of accidentally encouraging this through the order of the presentation, even though the approach clearly intends to assign personal responsibility where it is due (i.e. effective and feasible, as with air travel under certain conditions). The ECO_2-interview version showed participants that top-down social and/or technological changes were often, and in certain cases mostly, more effective than direct alterations in the household (although naturally such social changes could not be effected without the participation of a significant percentage of the population).[99] Then, in ranking actors' responsibility for national energy use, subjects apparently generally extended this lesson to the question of responsibility and put the onus of promoting at least technological advances on decision makers like corporate boards and politicians. If people view these decision makers as uninfluenceable – and all non-discretionary factors as truly and irrevocably non-discretionary on any policy-relevant time-scale – then we run the risk of Finger's and Meijnders' (1994) (worst-case) scenarios: High environmental consciousness, problem recognition and concern among the public, and a high level of recognition of institutional and technological constraints, but also a strong feeling of individual powerlessness and helplessness to do anything about them.[100] One consequence is a circle of information gathering as fatalistic confirmation or emotional solace, and because no other avenue is open: "Informing householders *as a substitute for* sustainable energy

[99] See the text and summarizing boxes in section 8.4.4 of Chapter 4.

[100] Futurist Joseph Coates says that our capability to anticipate and influence the future is undoubtedly constrained. But if we choose to remain *totally* limited, then we get exactly what we deserve: the worst of all possible worlds.

consumption"[101] because apparently learning about environmental threats is better than ignorance of them, even when they are ineluctable. Other possible end results are cynicism, apathy[102] or disengagement ("shutting out and tuning out"). As noted in Chapter 3, this danger is partly a product of the withering of Western democratic electorates' political participation. And matters are made worse if such energy programs accidentally teach users to abdicate what individual responsibility they took previously for their domestic consumption decisions – if, for example, interview participants leave with a highly technological optimist message and forget the lessons of the individual and Swiss aggregate graphs. After all, depending on the person, individual conservation efforts are important and necessary in certain sectors or under certain circumstances, and definitely vital in the aggregate to confront demand escalation and rebound.

This is where the notion of consumer involvement and participation, and its encouragement through "consumer-oriented monitoring" (e.g. counter-surveillance of producers) takes on importance as a form of democratic rejuvenation. The ECO_2 interviews only started explicitly to assess participants' views of possibilities here near the end of the interview. Views on the scope for consumer-citizen involvement were mixed and not uniformly encouraging. The balance of opinion was towards a limited scope or no individual influence at all, and a limited role for (individual) consumers in any case, which is surprising for such a market-oriented society.[103] Systematically exploring this issue, using statistically generalizable sampling of the population, is important for any future work of this sort and would be a logical extension of a (shortened) session with the energy accounting software.

4. SUCCESS IN RISK COMMUNICATION

According to Renn (1991b) success in risk communication (RC) depends partly on paying attention to the fact that "individuals as well as social units

[101] A play on the title of [Goldblatt 2002].

[102] Renn notes that personal apathy is one of the possible secondary effects of risk communication [Renn 1991a].

[103] Similar results were recently found in a larger study using focus groups to assess the market potential for green electricity, especially hydropower, in Switzerland [Spreng 2001]. The percentage of domestic customers who actually pay 20% more for environmentally produced electricity is low; the actualised green consumer niche is much smaller than that predicted by marketing studies, although it is expected to increase with the ease of switching providers in liberalized power markets.

make use of a complex variety of internal and external cues to process messages and that the variation of one or two factors may only lead to marginal changes in the outcome" [Renn 1991b]. By combining personal and aggregate/long-term program modules, and incorporating social, technological, and institutional factors, we tried to provide a measure of comprehensiveness and a large range of changeable parameters to achieve a richer variety of outcomes. The diversity and richness of responses and comments across subjects seems to point to success on this score. Using the program with people one-on-one allowed for highly individualized outcomes and tailored learning experiences, depending on the interviewee's current energy profile, her preferences for change, and her choice of scenario parameters according to her vision of future national developments. However, by experimental standards, changing so many variables, even when done stepwise and cumulatively, may remove an element of control seen as important for validating results. Since we did not set out to conduct a rigorously controlled experiment but rather a hybrid pilot study, this is not a major flaw.

However, from some risk communications perspectives, the interview sessions built in serious limitations, and the application of their results is therefore also limited. Kasperson warns against RC decomposition that assumes messages can be separated from their context of relationships and social interactions in general. The peculiarities of participants, their relationships, and their interactions can and maybe should be an important determinant of which information is exchanged, how it is to be interpreted, and what the continuation of the communication process will be like [Kasperson 1991]. Yet, the caveats at the conclusion of Chapter 3 warned against taking this so far as to elevate process over scientific content and thereby muddle the transdisciplinary aim. In the interview sessions, the balance between expert and lay input into the process was naturally somewhat weighted towards the expert side, since the generation of energy profiles and their manipulation relies so heavily on the use of model-based software. Lay input increased as the interviews proceeded to the questions of how people interpret and what use they make of the knowledge and insights they had gathered earlier. Again, concrete political applications, explored only tentatively near the end of the interviews, lend themselves to more social and participatory settings like focus groups, and these should be pursued in future extensions.

As expected, the combination of Energy-Revealing and Social-Revealing approaches allowed risk communication to proceed on several levels. The interviews operated on at least levels two and three, to use Renn's (1991b) categorizations (Chapter 3, section 3.2.3). Level two's reflective discourse started with the assumption of serious environmental risks, made clear in the response framework explained in texts offered to participants at the beginning of the interview, and proceeded to involve a variety of broadly construed risk management options. The more interesting achievement is the discussions the interview session generated on level three, which involves the broadest discourse on values, lifestyles, and technology's role in society, among other things. While the interviewees' views on these naturally varied, and values were not often explicitly invoked, the software program stimulated thoughts on the energy implications of different patterns of social and technological development in Switzerland. The juxtaposition of technological (Type I) and social (Type II) parameters (interview step 6) almost forced a confrontation with some elements from level three. And most participants, even orthodox schooled economists and technological optimists, found the program's demonstration of the importance of social and lifestyle factors compelling (or even eye-opening) and were willing to engage the topic on these terms. The ECO_2-interview version, informed by the Social-Revealing approach, has contributed to the development of a macro-sociological framework and risk communications medium suitable for Renn's third-level debate.

The interviews also probably induced both peripheral and central interpretation of the risk communication messages, with an emphasis on the peripheral. The issue of holistic trust in the model behind the software came up several times in the interviews (Chapter 4, section 8.8.2) and a few subjects noted that the majority of their conclusions from the program depended on their trust in the embedded model. One subject challenged the breadth of the technological vision embodied in the (highest modeled levels represented by the top drop-down choices for) Type I parameters and said our tailor-made scenarios represented only a small subset of future possibilities. As noted, others challenged this or that statistic that clashed with their notions of what was normal for their social circles in Switzerland.

Interestingly, some lay subjects challenged various aspects of the data or model they found incredible based on their own logic or experience; many would have been satisfied with corrections to these aspects and did not dismiss the entire model or conclusions they drew that relied on other factors unrelated to those they found troublesome. This suggests that when given access to certain model parameters and assumptions, if not the model's mathematical workings, targets of energy risk communication may react with more nuanced responses than Renn's standard three choices of accept (submit to cost-benefit tradeoffs), reject completely (and demand zero-risk), or remain in doubt and use peripheral signals rather than factual/logical reasoning to decide whether or not to accept the risk communicator's message. Rather, some of the public may be willing to debate the risk communicator on one or another input or conclusion that they can relate to, such as levels of activity or costs. This could be the equivalent of public input into the risk analysis process that, in a regulatory risk assessment setting, takes place through more formalized procedures but often produces similar results.

Chapter 3 suggested that the peripheral (and Social-Revealing) route might hold the potential to engage different audiences on the question of energy consumption, not just the environmentally concerned. By keeping mention of specific environmental details to a minimum (largely to the explanatory material in the beginning and the re-invoking of the 2000 watt per capita environmental threshold at the interview's end) and instead working with ordinary terms involving social practices, technology, and social or demographic activity levels on the input side and energy terms on the output side, the sessions sought to engage participants on the conservation issue from a broader range of motivations. A few, for instance, saw the ECO_2's personal module potentially useful to save money in the home. For the national component, some seemed to view the projected increase in Swiss energy use as economically troubling, while at least a few who assumed an energy coupling with GNP saw it as expected and positive. With only relatively few (non-representative) subjects, it cannot be clearly assessed how useful such a tool would be to engage people in energy consumption out of a wide range of concerns. The recent new oil price shocks and fears of a wider geopolitical destabilization as terrorism and infrastructure sabotage increase in the Middle Eastern oil-producing states themselves are drawing more attention to energy issues. National security and economic concerns may again join environmental concerns as the main draws for consideration of energy production and consumption. General consumption or consumerism may attract a somewhat larger constituency than purely energy consumption but is still more a concern for the fringe than the mainstream.

5. SUGGESTIONS FOR FURTHER RESEARCH

The interviews showed that the software has considerable pedagogical value, even though it was designed for experimental purposes to point the way towards full-fledged packages that could incorporate those features of the approach that proved themselves successful. Chapter 4 reviewed interviewees' suggestions for improving specific design and functional elements of the ECO_2 software. Certain questions at the end of the interviews were meant to contribute to future, more contextual studies, for instance interview step 9, which asked subjects for ideas on how they and others could become involved in improving the energy performance of infrastructures, technologies, and/or social practices. As mentioned, some Dutch work is being developed that uses the social-practices approach involving lifestyles and infrastructural influences on householders' use of electricity [van den Burg 2001]. They also use energy accounting software developed for use in interviews. The focus, however, seems to be on the market-oriented and political details of (implementing) reduction strategies. The study thus represents a potential next step, albeit in a setting limited largely to direct electricity use.

In future work with the ECO_2–interview version, shortening and streamlining the presentation further (e.g. by focusing only on certain household sectors using direct energy) would help avoid accidental overemphasis of either technological or social factors. In a sector-limited treatment, however, while certain micro-macro comparisons are possible and desirable, overall scale issues would be harder to address (personal energy consumption could just be shunted from direct household uses into goods, leisure, transport, or other services not examined, canceling out reductions in the sectors of focus).

It was suggested in several places in Chapter 4 that future applications geared towards laypeople should tailor the personal data more specifically to relevant social groups. More (live) comparison of consumption with that of one's peer group and the sharing of experiences and savings success stories with peers may be advisable. Eco-team's group meetings may be that program's most important source of support and incentives to make enduring lifestyles changes. This feature was notably absent in the ECO_2 interview settings. The next step involving political contextualization should then be in a (focus) group setting. The ideal order may be a shorter individual, interview-guided session using ECO_2 and then a group meeting applying the lessons to end-user involvement in greening the infrastructures of consumption.

"Risk communication is part of a complex communications web in which various groups and cultures possess varying perceptions, values, and interpretations" [Kasperson 1991]. Our empirical study was limited by subject selection and analytical methods; in particular, the influence of different values (especially among cultural sub-groups), while recognized from the start as being important to the question set, was jettisoned early on in the interview planning process as infeasible with this software tool in the chosen communications setting. The influence of diverse and competing values could also be developed in future studies along these lines.

Current research into similar but more ambitious sustainability tools shows that not only is it possible to stimulate dynamic interplay between the user and expert knowledge embedded in the tools, but that put into action at schools and homes, the tools can help people move from education and knowledge to participation in social change. Georgia Basin (GB) QUEST, a computer simulation game developed by University of British Columbia's Sustainable Development Research Institute as part of its Georgia Basin Futures Project,[104] generates scenarios of sustainable futures based on users' choices of a wide variety of social and technological parameters and assumptions. GB QUEST's underlying conceptual framework is built around the conventional triple sustainability imperatives and uses the strategies of dematerialization and resocialization. Applications and similar projects are in development or planned for cities around the world (Goldblatt 2003).

Grappling with the difficulties of breaking out of the North's unsustainable consumption trajectory, Chapter 2 (section 4.1) described this conundrum: "Whence comes the political stimulus for the necessary restraint-enabling institutional and structural changes, when only these changes would bring home the timely reality of personal and organizational threat?" The theoretical elaboration of politico-economic requisites for sustainable consumption is a job for ecological economists and political scientists. Princen's answer might be independently to convince people of the need and practicality of changing economic institutions so as to restore signals for restraint, among both consumers and important players in the globalized and dispersed production-consumption chains. This would go considerably beyond conventional market mechanisms like energy taxes or emissions quotas, but it is based on the same reasoning that the system should provide the right incentives for players at all levels and that appeals for individuals to go against the grain are nearly useless. Looking back over the interview sessions and asking how close the subjects' responses came to giving a comprehensive description of the whole consumption problem – in other words, how close the subjects' program and interview experience came to the (expert) account in Chapter 2 – we note that such structural economic elements usually played a relatively small role in lay accounts. This is not surprising considering that the modeled non-discretionary influences did not specifically include this type of element. An (Ecological) Economic-Revealing approach to consumption would require a different set, or extended version, of parameters in energy software beyond the technological and social parameters modeled in the ECO_2 program. Applying such an idea to the end-user information approach for sustainable consumption (just as social and technology theory was applied here to consumer monitoring and information), one could imagine for example its embodiment in an analogous information-modeling tool.

[104] www.basinfutures.net

Tailoring this tool to the end-user, however, in the hope of consumer involvement for change is likely to be conceptually inaccessible and fruitless. Here there may be no substitute for organized pressure on the new nodes of power both upstream (e.g. institutional investors and speculators) and downstream (e.g. brand-name merchandisers, marketers, and advertisers) in global commodity chains [Conca 2001]. This would reinforce a consumer-oriented Social-Revealing approach, since consumers' practices and attitudes are increasingly influenced by financial and corporate decision makers such as these.

In the meantime, the present application is rich enough to use productively with energy policy-makers themselves. The program's combination of the national scale with the household module could give policy-makers insights as both professionals and householders. Many planners and policy-makers would appreciate the long-term scenario capabilities of the national screen and could learn much from experimenting with the interaction among the levels and time-scales. Seeing how technology choices lock in future energy use or how aggregate social developments overwhelm individual efforts or efficiency improvements is important and new for many policy-makers, and lessons they draw have the potential to be implemented directly.

6. SOME OPEN QUESTIONS

Here are some additional relevant open questions for the sustainable consumption agenda [Spaargaren 2003] that could be explored in future applications of this research:

1. What are the new social relations that accompany the socio-technical innovations implied or required by the ecological modernization of consumption?
2. Does the "greening" of activity sectors and social practices like transportation or diet favor expert knowledge over the participation in and influence of laypeople on providers, producers, infrastructure, and technology development? The interviews sought to stimulate lay involvement but went about it in a partially expert manner (cf. section 4). A related item is evaluating and overcoming cultural resistance to the use of the knowledge gained by lay users of such programs as ECO_2. We can imagine that necessary institutional change may also have to extend to institutions that hinder people's use of the sort of knowledge thus imparted, either in the home or externally by consumer or political involvement. What might these institutional barriers be?

3. A cluster of basic open questions for environmental sociologists researching the greening of institutions is the following: How much lay involvement, co-provision, or co-production of energy related systems, artifacts, and other structural elements can be expected or encouraged (or is already observable), and how much rather is democratic wishful thinking on the part of green social scientists? Can consumer-oriented monitoring change infrastructures of consumption (individual or collective consumption practices and/or those of providers or producers) in the absence of regulation or other behavior-forcing mechanisms [van den Burg 2001]? That is, how much can even enhanced voluntary consumer information approaches be relied upon? Further, how much of what researchers have seen is idiosyncratic to the cultural and political landscape of their research areas (e.g. the Netherlands) and how much is truly extendable in theory or practice to other industrialized or even industrializing nations? And, if consumer-citizen involvement can be anticipated with confidence in a certain sector and region, what sort of relations or institutions can be looked towards to encourage an environmentally favorable outcome? Consumer-oriented monitoring or end-user involvement does not necessarily or even predominantly imply pursuit of an environmentalist agenda, as sometimes seems to be the implicit assumption of its advocates. The tendency to deify consumer sovereignty and neglect the problematic of demand escalation and scale issues (understandably to preserve amity with co-researchers, funders, and policy makers) threatens to distance the ecological modernization agenda from certain fundamental reforms necessary for sustainable consumption.

This segues into some final remarks on extensions to and departures from ecological modernization.

7. DIVERGENCE FROM THE PREVAILING ENVIRONMENTAL FRAMEWORK

This study has often pointed up the apparent affinity of its agenda, conceptual framework, and experimental monitoring approach with those of ecological modernization (EM) theory, especially the modern Dutch variety represented by Spaargaren and Mol. Chapters 3 and 4 make a contribution to certain fundamental issues and mechanics of monitoring for the ecological modernization of consumption, even though they did not initially set out to do so: our approach came out of the even broader consideration of the consumption problematique presented in Chapters 1 and 2. Moreover, EM has rapidly become influential in environmental social science and policy, at least in areas of Western Europe, and the term is even used more broadly to characterize strategic environmental management, industrial ecology, or environmental improvement in general [Buttel 2000]. It is therefore appropriate to end this reflection on lessons learned by considering how far EM goes in dealing with the problem as laid out in the first two chapters of this book, and what it still leaves to be desired.

Ecological modernization theory itself (like sustainable consumption to some extent) arose partly in response to shortcomings of sustainable development, formulated originally with regard to policies toward the South and from experiences there, in addressing Northern environmental problems. EM is also to some extent a critique and response to radical environmentalism ("countermodernity"). EM theory is young, still evolving, grounded in historical and political reality, not canonical, and variegated. Generally, EM is optimistic about the malleability of the institutions and technological capabilities of industrial capitalism [Buttel 2000]. Spaargaren and Mol's sociological conception of EM hypothesizes that not only is capitalism sufficiently flexible institutionally to permit movement in the direction of "'sustainable capitalism', but its imperative of competition among capitals can – under certain conditions – be harnessed to achieve pollution-prevention eco-efficiencies within the production process and ultimately within consumption processes as well" ([Spaargaren 1996] as cited in [Buttel 2000]). "The environment becomes relatively independent (now from the economy), ultimately having as a consequence that a capitalist or rather market-based system of production and consumption does not necessarily contradict significant environmental improvements and reforms in any fundamental way" [Mol 2000]. Buttel identifies some potential shortcomings of current EM as a social theory, among them an overconfidence in the "transformative potentials" of modern capitalism and a relative neglect of broader concerns about aggregate resource consumption and its environmental impacts [Buttel 2000]. Princen, Conca, and many ecological economists would likely concur and claim that the reforms needed in global capitalism may amount to more than evolutionary institutional change.

"In the end, the empirical question will of course remain whether these radical environmental reforms {in the EM vein} will be sufficient to deal with the – to a large extent socially constructed – criterion of sustainability" [Mol 2000]. It is safe to say that a majority of natural and physical scientists working in the broad field of environment and development, however they construct sustainability criteria, would answer the question "no, they are insufficient now and will prove insufficient at least in the immediate future." And to the extent that they are biogeophysically constructed, sustainability criteria show that whatever case studies ecological modernizationists adduce as evidence of environmentally oriented institutional transformations thus far, global environmental results are sorely lacking. Most environmental "state" indicators and trends worldwide are, to say the least, disheartening.

EM champions the emergence of a separate ecological rationality to judge individuals' and institutions' actions in which the "rationally calculating citizen will be just as keen on avoiding environmental risks ... as she is on realising economic benefits or enhancing status" [Spaargaren 2000b]. But it does not seem to advocate the sort of structural reforms which Princen and others say are necessary to restore feedback concerning the ecological risk of their decisions to actors along the consumption-production chain, instead apparently viewing the sort of evolutionary economic changes already occurring as sufficient. Thus, there may be an agreement on ends but a disagreement on necessary means.

The previous section noted the tendency towards over-emphasizing consumer sovereignty and keeping the lifestyles, wishes, and demands of consumers sacrosanct. Yet, the constant and now globalized escalation and expansion of those wishes and demands is a significant driver for environmental degradation. Wilk's social reconversion of needs to wants and Princen's restoration of restraint then seem to be extensions of or departures from EM. Following the EM orientation in practice also seems to tend towards, but in theory certainly does not require, neglecting the role of responsibility on the "human agency" end of the discretion continuum.

Dealing with scale and sufficiency in the (global) macroeconomy likewise seems to be out of the scope of EM reforms. EM deals with the macro-level effects of technology on the institutional side, but is it willing for instance to educate about the effects of overconsumption through its efforts in consumer-oriented monitoring?

Along these lines, a bold engineering professor and political scientist recently wrote: "How might an engineering educator raise such concerns {about environmental dangers of overconsumption}? The key is not to focus on a toaster, or even a power plant, because the problem of overconsumption arises largely from the overall set of toasters and power plants rather than from any one of them individually ... quantity and variety would still be excessive for a world of 8-12 billion quasi-Americans. *So a class somehow needs to tackle the generic problem, or at least a piece of it that illustrates the larger situation*" [Swearengen 2003 (emphasis added)]. The ECO_2 interviews incorporated scale when they introduced the 2000 watt per capita target. But confronting limits is not likely to be a popular avenue for monitoring for the ecological modernization of consumption, since its implications are too upsetting to the assumptions of the prevailing economic order.

This argues again for an Economic-Revealing approach to analyzing consumption chains. As noted, such an approach may not be suitable for a consumer-oriented monitoring tool like ECO_2. In any realized form, in its necessary embrace of ecological economics, an Economic-Revealing analytic approach would probably represent a departure from ecological modernization theory. But environmental theories and broader policies will have to veer from convention and embrace elements in Economic- as well as Social- and Environment-Revealing agendas if they hope to improve future prospects for global environmental sustainability significantly.

References

Aebischer, Bernard (1998), "Energy Consumption Indicators in Switzerland 1960-1985/97: Highlights of the Intermediate Report", Energy Analysis Research Group, Zurich: Swiss Federal Institute of Technology.

[Agenda 21 1992] http://www.un.org/esa/sustdev/agenda21.htm

Aho, Leena (1998), "The Theoretical and Practical Basis for Environmental Teaching and Learning for Sustainable Development", in: Filho, Walter Leal and Ahlberg, Mauri (eds.), *Environmental Education for Sustainability: Good Environment, Good Life*, Frankfurt am Main: Peter Lang GmbH.

Ajzen, I. and Madden, T.J. (1986), "Prediction of Goal-Directed Behaviour: Attitudes, Intentions, and Perceived Behavioural Control", *Journal of Experimental Social Psychology*, vol. 22.

Bernard, M.J. III, McBride, J.R., Desmond, D.J., Collins, N.E. (1988), "Events – The Third Variable in Daily Household Energy Consumption", *Proceedings of the American Council for an Energy-Efficient Economy*, Nov. 11-14, Washington D.C.: ACEEE Press.

Bijker, Wiebe E. (1995), *Of Bicycles, Bakelites, and Bulbs: Toward a Theory of Socio-technical Change*, Cambridge Mass.: MIT Press.

Bolscho, Dietmar; Eulefeld, Guenter; Seybold, Hansjörg (1994), *Bildung und Ausbildung im Umweltschutz*, Bonn: Economica Verlag.

Bovay, Claude; Campiche, Roland J.; Hainard, Francois; Kaiser, Helmut; Pedrazzini, Yves; Ruh, Hans; Spescha, Plasch (1987), *L'énergie au quotidien: Aspects sociologiques et éthiques de la consommation d'énergie*, Geneva: Editions Labor et Fides.

Bovay, Claude; Campiche, Roland J.; Germann, Hans Ulrich; Hainard, Francois; Kaiser, Helmut; Pedrazzini, Yves; Ruh, Hans; Spescha, Plasch (1989), *Energie im Alltag: Soziologische und ethische Aspekte des Energieverbrauchs*, Zürich: Theologischer Verlag.

Brooks, David (2001), "One Nation, Slightly Divisible", *The Atlantic Monthly*, vol. 288, no. 5.

Brower, Michael and Leon, Warren (1999), *The Consumer's Guide to Effective Environmental Choices: Practical Advice from the Union of Concerned Scientists*, New York: Three Rivers Press.

Buttel, Frederick H. (2000), "Ecological Modernization as Social Theory", *Geoforum*, vol. 31.

Chappells, Heather; Klintman, Mikael; Linden, Anna-Lisa; Shove, Elizabeth; Spaargaren, Gert; van Vliet, Bas (2000), *Domestic Consumption Utility Services and the Environment: Final Domus Report*, Universities of Lancaster, Wageningen, and Lund.

Charkiewicz, Eva; van Bennekom, Sander; Young, Alex (2001), *Transitions to Sustainable Production and Consumption: Concepts, Policies and Actions*, The Hague: Tools for Transition.

Cogoy, Mario (1995), "Market and Non-Market Determinants of Private Consumption and their Impacts on the Environment", *Ecological Economics,* vol. 13, no. 3.

Cohen, Maurie J. (2000), "Ecological Modernisation, Environmental Knowledge and National Character: A Preliminary Analysis of the Netherlands", in: Mol, Arthur P. J. and Sonnenfeld, David A. (eds.), *Ecological Modernisation around the World; Perspectives and Critical Debates*, London: Cass.

Cohen, Marie J. (2001), "The Emergent Environmental Policy Discourse on Sustainable Consumption", in: Cohen, Marie J. and Murphy, Joseph (eds.), *Exploring Sustainable Consumption: Environmental Policy and the Social Sciences*, Amsterdam: Pergamon.

Coley, D.A., Goodliffe, E., Macdiarmid, J. (1998), "The Embodied Energy of Food: The Role of Diet", *Energy Policy*, vol. 26, no. 6.

Commoner, B., Corr, M., Stamler, P. (1971), "The Causes of Pollution", *Environment*, April.

Conca, Ken (2001), "Consumption and Environment in a Global Economy", *Global Environmental Politics*, vol. 1, no. 3.

Connors, Stephen R. (1998), "Issues in Energy and Sustainable Development", *AGS Mapping Project White Paper*, Cambridge: MIT Energy Laboratory.

Crocker, David (1994), "Consumption, Well-Being, and Virtue", in: Goodwin, Neva; Ackerman, Frank; Kiron, David (eds.) (1997) *The Consumer Society*, Washington D.C.: Island Press.

Crocker, David and Linden, Toby (eds.) (1998), *Ethics of Consumption: The Good Life, Justice, and Global Stewardship,* Lanham MD: Rowman & Littlefield.

Dahinden, Urs and Dürrenberger, Gregor (1997), "Public Participation in Energy Policy: Results from Focus Groups", in: Renn Ortwin (ed.), *Risk Perception and Communication in Europe*, vol. 2 of *Risk Analysis and Management in a Global Economy*, Baden-Württemberg: Center of Technology Assessment.

Dahinden, Urs (2000), *Demokratisierung der Umweltpolitik: Ökologische Steuern im Urteil von Bürgerinnen und Bürgern*, Baden-Baden: Nomos Verlagsgesellschaft.

Dahl, L. (1995), "Towards Indicators of Sustainability," SCOPE Scientific Workshop on Indicators of Sustainable Development.

Daly, Herman E. and Cobb, John B. (1994), *For the Common Good: Redirecting the Economy toward Community, the Environment, and a Sustainable Future*, Boston: Beacon Press.

Daly, Herman (1996a), *Beyond Growth: The Economics of Sustainable Development*, Boston: Beacon Press.

Daly, Herman (1996b), "Consumption: Value Added, Physical Transformation, and Welfare", in: Goodwin Neva, Frank Ackerman, David Kiron, (eds.) (1997), *The Consumer Society*, Washington D.C.: Island Press.

Daly, Herman E. (1998), "Reply to Mark Sagoff's 'Carrying Capacity and Ecological Economics'", in: Crocker, David and Linden, Toby (eds.) (1998), *Ethics of Consumption: The Good Life, Justice, and Global Stewardship,* Lanham MD: Rowman & Littlefield.

Dholakia, Ruby R. and Dholakia, Nikhilesh (1983), "From Social Psychology to Political Economy: A Model of Energy Use Behavior", *Journal of Economic Psychology*, vol. 3.

Dobson, Andrew (1990), *Green Political Thought: An Introduction*, London: Unwin Hyman.

Douglas, M. and Isherwood, B. (1979), *The World of Goods: Towards an Anthropology of Consumption*, London: Allen Lane.

Durning, Alan (1991), "Asking How Much is Enough", in: Goodwin, Neva; Ackerman, Frank; Kiron, David (eds.) (1997), *The Consumer Society*, Washington D.C.: Island Press.

Easterlin, Richard (1995), "Will Raising the Incomes of All Increase the Happiness of All?", in: Goodwin, Neva; Ackerman, Frank; Kiron, David (eds.) (1997), *The Consumer Society*, Washington D.C.: Island Press.

Ehrlich, Paul and Holdren, John (1972), "Review of The Closing Circle", *Environment*, April.

Ekins, Paul (1993), "'Limits to Growth' and 'Sustainable Development': Grappling with Ecological Realities", *Ecological Economics*, vol. 8, no. 3.

Farhar, Barbara C. (1994), "Trends in US Public Perceptions and Preferences on Energy and Environmental Policy", *Annual Review of Energy and Environment*, vol. 19.

Filho, Walter Leal and Ahlberg, Mauri (eds.) (1998), *Environmental Education for Sustainability: Good Environment, Good Life*, Frankfurt am Main: Peter Lang GmbH.

Finger, Matthias (1994), "From Knowledge to Action? Exploring the Relationships between Environmental Experiences, Learning, and Behavior", *Journal of Social Issues*, vol. 50, no. 3.

Fischer-Kowalski, Marina (1998), "Society's Metabolism: Intellectual History of Materials Flow Analysis, Part I, 1960-1970", *Journal of Industrial Ecology*, vol. 2, no. 1.

Flury, Manuel (ed.) (1995), *Bevölkerungsentwicklung, Lebensstil und Umweltverantwortung*, Bern: Interfakultäre Koordinationsstelle für Allgemeine Ökologie.

Foddy, William (1993), *Constructing Questions for Interviews and Questionnaires: Theory and Practice in Social Research*, Cambridge: Cambridge University Press.

Galbraith, John Kenneth (1997), "Foreword", in: Goodwin, Neva; Ackerman, Frank; Kiron, David (eds.), *The Consumer Society*, Washington D.C.: Island Press.

Garcia, Ernest and Duart, P. (1998), "Consumo y sustentabilidad en la Comunidad Valenciana", *Revista Internacional de Sociología* (RIS), tercera época, no. 19-20.

Gatersleben, Birgitta and Vlek, Charles A.J. (1998), "Household Consumption, Quality of Life, and Environmental Impacts: A Psychological Perspective and Empirical Study", in: Noorman, Klaas Jan and Uiterkamp, Ton Schoot (eds.), *Green Households? Domestic Consumers, Environment, & Sustainability*, London: Earthscan.

Georg, Susse (1999), "The Social Shaping of Household Consumption", *Ecological Economics*, vol. 28, no. 3.

Georgescu-Roegen, Nicholas (1971), *The Entropy Law and the Economic Process*, Cambridge, Mass.: Harvard University Press.

Giddens, Anthony (1984), *The Constitution of Society*, Berkeley: University of California Press.

Giddens, Anthony (1990), *The Consequences of Modernity*, Cambridge: Polity Press.

Goldblatt, David L. (1993), *The Defeat of S 1220's CAFE Title: Examinations of the Senate's Failure to Raise Automotive Fuel Economy Standards in 1991*, Yale University F&ES Master's Thesis (unpublished), New Haven.

Goldblatt, David L. (1998), "Case Study: Analysis of the Energy Impact of Electric-assist Bicycles in India and China", Energy Analysis Research Group, Zurich: Swiss Federal Institute of Technology.

Goldblatt, David L. (2002), *Personal vs. Socio-technical Change: Informing and Involving Householders for Sustainable Energy Consumption*, Zurich: ETH dissertation, no. 14758.

Goldblatt, David L. (2003), "A dynamic structuration approach to information for end-user energy conservation", *Time to Turn Down Energy Demand: Proceedings of the 2003*

ECEEE Summer Study, Prague/Copenhagen: European Council for an Energy-Efficient Economy.

Goldblatt, David L.; Dürrenberger, Gregor; Hartmann, Christoph (2005), "Combining interviewing and modeling for end-user energy conservation", *Energy Policy, vol. 33, no. 2.*

Goldemberg, José et. al. (1987), *Energy for a Sustainable World,* Washington D.C.: World Resources Institute.

Goldemberg, José [2000], *World Energy Assessment: Energy and the Challenge of Sustainability*, New York: UNDP.

Goodland, Robert; Daly, Herman; El Serafy, Salah (eds.) (1992), *Population, Technology, and Lifestyle: the Transition to Sustainability,* Washington D.C.: Island Press.

Goodwin, Neva; Ackerman, Frank; Kiron, David (eds.) (1997), *The Consumer Society*, Washington D.C.: Island Press.

Goodwin, Neva (1997a), "Volume Introduction", in: Goodwin, Neva; Ackerman, Frank; Kiron, David (eds.), *The Consumer Society*, Washington D.C.: Island Press.

Goodwin, Neva (1997b), "Overview Essay Part I", in: Goodwin, Neva; Ackerman, Frank; Kiron, David (eds.), *The Consumer Society*, Washington D.C.: Island Press.

Goodwin, Neva (1997c), "Overview Essay Part X", in: Goodwin, Neva; Ackerman, Frank; Kiron, David (eds.), *The Consumer Society*, Washington D.C.: Island Press.

Haavelmo, T. and Hansen, S. (1992), "On the Strategy of Trying to Reduce Economic Inequality by Expanding the Scale of Human Activity", in: Goodland, Robert; Daly, Herman; El Serafy, Salah (eds.), *Population, Technology, and Lifestyle: the Transition to Sustainability,* Washington D.C.: Island Press.

Harris, Jonathan (1997), "Overview Essay Part VIII", in: Goodwin, Neva; Ackerman, Frank; Kiron, David (eds.), *The Consumer Society*, Washington D.C: Island Press.

Harrison, Carol and Davies, Gail (1998), "Lifestyles & the Environment", *Environment and Sustainability Desk Study*, London: University College.

Hawken, Paul; Lovins, Amory; Lovins, L. Hunter (1999), *Natural Capitalism: Creating the Next Industrial Revolution,* Boston: Little, Brown & Co.

Herendeen, Robert A. (1998), *Ecological Numeracy: Quantitative Analysis of Environmental Issues*, New York: John Wiley & Sons.

Holdren, John; Daily, Gretchen; Ehrlich, Paul (1995), "The Meaning of Sustainability: Biogeophysical Aspects" In: Munasinghe, Mohan and Shearer, Walter (eds.), *Defining and Measuring Sustainability: Biogeophysical Foundations,* Washington D.C.: United Nations University/ World Bank.

Honkasalo, Antero (1998), "Entropy, Exergy and Steady-State Economy", *Sustainable Development,* vol. 5, no. 6.

Hughes, Thomas (December 11, 1997), Lecture given at the Swiss Federal Institute of Technology, Zurich.

Hurtado, Maria Elena (1996), "Sustained Advertising Promoting Unsustainable Consumption", *Briefing Paper: Consumer Unity & Trust Society*, no. 2.

Imboden, Dieter M. and Jaeger, Carlo C. (1999), "Towards a Sustainable Energy Future", in: *Energy: The Next Fifty Years,* Paris: OECD.

Immerwahr, John (1999), "Waiting for a Signal: Public Attitudes toward Global Warming, the Environment and Geophysical Research," http://www.agu.org/sci_soc/attitude_study.html

Irwin, Alan and Wynne, Brian (eds.) (1996a), *Misunderstanding Science? The Public Reconstruction of Science and Technology*, Cambridge: Cambridge University Press.

Irwin, Alan (1996b), "Conclusions", in: Irwin, Alan and Wynne, Brian (eds.), *Misunderstanding Science? The Public Reconstruction of Science and Technology*, Cambridge: Cambridge University Press.

Jackson, Tim and Marks, Nic (1999), "Consumption, Sustainable Welfare and Human Needs – with Reference to UK Expenditure Patterns between 1954 and 1994", *Ecological Economics,* vol. 28, no. 3.

Jaeger, Carlo (1994), *Taming the Dragon: Transforming Economic Institutions in the Face of Global Change*, Yverdon: Gordon and Breach.

Jochem, Eberhard; Satheye, Jayant; Bouille, Daniel (eds.) (2000), *Society, Behaviour, and Climate Change Mitigation*, Dordrecht: Kluwer Academic Publishers.

Kaiser, Florian (1998a), "A General Measure of Ecological Behavior", *Journal of Applied Social Psychology*, vol. 28, no. 5.

Kaiser, Florian (1998b), "Person und Situation als Determinanten unterschiedlicher Aspekte ökologischen Verhaltens", *Umweltpsychologie*, vol. 2, no. 1.

Kaiser, Florian; Wolfing, Sybille; Fuhrer, Urs (1999a), "Environmental Attitude and Ecological Behaviour", *Journal of Environmental Psychology*, vol. 19.

Kaiser, Florian and Fuhrer, Urs (1999b), "Ecological Behavior's Dependency on Different Forms of Knowledge", *Applied Psychology: An International Review* (forthcoming).

Kant, A.D. and Midden, C.J.H. (1995), *Energy and the Public – Phase 2 Public Communication in Energy Issues: Psychological Evaluation Models*, Petten: Netherlands Energy Research Foundation ECN.

Kapyla, Markku (1998), "Environmental Education and Implicit Meanings", in: Filho, Walter Leal and Ahlberg, Mauri (eds.), *Environmental Education for Sustainability: Good Environment, Good Life*, Frankfurt am Main: Peter Lang GmbH.

Kasperson, Roger E. (1986), "Six Propositions for Public Participation and Their Relevance for Risk Communication", *Risk Analysis,* vol. 6, no. 3.

Kasperson, Roger E. and Stallen, Pieter Jan M. (eds.) (1991), *Communicating Risks to the Public: International Perspectives,* Dordrecht: Kluwer Academic Publishers.

Kaufmann-Hayoz, Ruth et. al. (eds.) (1999), *... man kann ja nicht einfach aussteigen: Kinder und Jugendliche zwischen Umweltangst und Konsumlust*, Zürich: Vdf, Hochschulverlag an der ETH Zürich.

Kempton, Willett (1991), "Lay Perspectives on Global Climate Change", *Global Environmental Change, Human and Policy Dimensions*, vol. 1, no. 3.

Kempton, Willett (1993), "Will Public Environmental Concern Lead to Action on Global Warming?", *Annual Review of Energy and Environment*, vol. 18.

Kempton, Willett and Payne, C. (1997), "Cultural and Social Evolutionary Determinants of Consumption", in: Stern, Paul C. et. al., *Environmentally Significant Consumption: Research Directions,* Washington, D.C.: National Academy Press.

Kiron, David (1997), "Overview Essay Part II", in: Goodwin, Neva; Ackerman, Frank; Kiron, David (eds.), *The Consumer Society*, Washington D.C.: Island Press.

Klaus, Eder; Lash, Scott; Strydom, Piet; Dobry, Michel; Brand, Karl-Werner (1995), "Framing and Communicating Environmental Issues: Summary Final Report", Cordis Document Library EV5V-CT92-0153, http://www.npk.gov.pl/cordis/www.cordis.lu/env/src/p-rep6.htm

Kok, Marcel; Vermeulen, W.J.V.; Faaij, A.P.C.; de Jager, D. (eds.) (2002), *Global Warming and Social Innovation: The Challenge of a Climate-Neutral Society*, London: Earthscan.

Kull, Stephen (2000), "Americans on the Global Warming Treaty", http://www.pipa.org/OnlineReports/GlobalWarming/buenos_aires.htm

de Laat, Sebastian (1997), "Reversing lifestyles: Future Energy Technologies as a Focus for Analysing Future Energy Behaviour", *The Energy Efficiency Challenge: Proceedings of the 1997 ECEEE Summer Study*, vol. 2, no. 151, Prague/Copenhagen: European Council for an Energy-Efficient Economy.

Lasch, Christopher (1991), *The True and Only Heaven: Progress and its Critics*, New York: Norton.

Leiss, William (1978), "The Limits to Satisfaction: Examination", in: Goodwin, Neva; Ackerman, Frank; Kiron, David (eds.) (1997), *The Consumer Society*, Washington D.C.: Island Press.

Lemert, Charles (ed.) (1999), *Social Theory: The Multicultural and Classic Readings*, 2nd ed., Boulder: Westview.

Ligteringen, Josee (1998), "The Effects of Public Policies on Household Metabolism", in: Noorman, Klaas Jan and Uiterkamp, A.J.M. Schoot (eds.), *Green Households? Domestic Consumers, Environment, & Sustainability,* London: Earthscan.

Linderhof, Vincent G.M and Kooreman, Peter (1998), "Economic Aspects of Household Metabolism", in: Noorman, Klaas Jan and Uiterkamp, A.J.M. Schoot (eds.), *Green Households? Domestic Consumers, Environment, & Sustainability,* London: Earthscan.

Lintott, John (1998), "Beyond the Economics of More: The Place of Consumption in Ecological Economics", *Ecological Economics,* vol. 25, no. 3.

Lovins, Amory (1977), *Soft Energy Paths: Toward a Durable Peace*, Cambridge Mass.: Ballinger.

Lutzenhiser, Loren (1993), "Social and Behavioral Aspects of Energy Use", *Annual Review of Energy and Environment,* vol. 18.

Lutzenhiser, Loren (1997), "Social Structure, Culture, and Technology: Modeling the Driving Forces of Household Energy Consumption", in: Stern P.C et. al., *Environmentally Significant Consumption: Research Directions,* Washington D.C.: National Academy Press.

Madden, M.J., Ellen, P.S., Ajzen, I. (1992), "A Comparison of the Theory of Planned Behavior and the Theory of Reasoned Action", *Personality and Social Psychology Bulletin,* vol. 18, no.1.

Maniates, Michael (2001), "Individualization: Plant a Tree, Buy a Bike, Save the World?", *Global Environmental Politics*, vol. 1, no. 3.

Manoochehri, John (2001), *Consumption Opportunities: Strategies for Change*, Geneva: UNEP.

Matthews, Emily and Hammond, Allen (1999), *Critical Consumption: Trends and Implications: Degrading Earth's Ecosystems*, Washington D.C.: World Resources Institute.

[Measuring Changes 1998] Division for Sustainable Development (1998), *Measuring Changes in Consumption and Production Patterns*, Department of Economic and Social Affairs, New York: United Nations.

Meijnders, Anneloes (1998), *Climate Change and Changing Attitudes: Effect of Negative Emotion on Information Processing*, Eindhoven: Eindhoven University of Technology.

Menanteau, Phillipe (1997), "Energy Efficiency Labeling for Appliances: A New Use of an Old Instrument in the French Region Nord-Pas de Calais", *The Energy Efficiency Challenge: Proceedings of the 1997 ECEEE Summer Study*, Prague/Copenhagen: European Council for an Energy-Efficient Economy.

Meppem, Tony and Bourke, Simon (1999), "Different Ways of Knowing: A Communicative Turn Toward Sustainability", *Ecological Economics,* vol. 30, no. 3.

Michaelis, Laurie (1999), "Policies for Sustainable Consumption: Research Issues", Oxford: Oxford Centre for the Environment, Ethics, and Society.

Michael, Mike (1996), "Ignoring Science: Discourses of Ignorance in the Public Understanding of Science", in: Alan, Irwin and Wynne, Brian (eds.), *Misunderstanding Science? The Public Reconstruction of Science and Technology*, Cambridge: Cambridge University Press.

Moezzi, Mithra (1998), "The Predicament of Efficiency", *ACEEE 1998 Summer Study Proceedings*, Washington D.C.: American Council for an Energy-Efficient Economy.

Mohler, Peter; Calvi, Gabriele; Ester, Peter; Jowell, Roger; Ward, Conor (1994), *Research into Environmental Attitudes and Perceptions (REAP): Summary Final Report*, Cordis Document Library EV5V-CT92-0172, http://www.zuma-mannheim.de/cooperation/en/compass/reap.htm.

Mol, Arthur P.J. (1995), *The Refinement of Production: Ecological Modernization Theory and the Chemical Industry*, Utrecht: Van Arkel.

Mol, Arthur P.J. and Spaargaren, Gert (2000), "Ecological Modernisation Theory in Debate: A Review", *Environmental Politics,* vol. 9, no.1.

[More with Less 1995] (1995), More with Less: Initiatives to Promote Sustainable Consumption, Commonwealth of Australia.

Munasinghe, Mohan and McNeely, J. (1995a), "Key Concepts and Terminology of Sustainable Development" in: Mohan, Munasinghe and Shearer, Walter (eds.), *Defining and Measuring Sustainability: Biogeophysical Foundations,* Washington D.C.: United Nations University/ World Bank.

Munasinghe Mohan and Shearer, Walter (eds.), (1995b), *Defining and Measuring Sustainability: Biogeophysical Foundations*, Washington D.C.: United Nations University/World Bank.

Myers, Norman and Kent, Jennifer (2004), *The New Consumers: The Influence of Affluence on the Environment*, Washington D.C.: Island Press.

[New Consumers 1999] (1999), "New Consumers: A Discussion with Dr. Norman Myers" (Electronic Conference), http://iisd.ca/susprod/newconsumers.htm.

Noorman, Klaas Jan and Uiterkamp, A.J.M. Schoot (eds.), (1998a), *Green Households? Domestic Consumers, Environment, & Sustainability,* London: Earthscan.

Noorman, Klaas Jan; Biesiot, Wouter; Uiterkamp, A.J.M. Schoot (1998b), "Household Metabolism in the Context of Sustainability and Environmental Quality", in: Noorman, Klaas Jan and Uiterkamp, A.J.M. Schoot (eds.), *Green Households? Domestic Consumers, Environment, & Sustainability,* London: Earthscan.

Ospelt, Christoph (1995), "Der direkte und der indirekte Energieverbrauch der Haushalte in der Schweiz: Konzept zur Berechnung unter Verwendung der Input-Output Analyse", Zurich: Swiss Federal Institute of Technology.

Pfister, Christian (ed.) (1996), *Das 1950er Syndrom: der Weg in die Konsumgesellschaft,* Bern: Haupt.

Pfister, Christian et. al. (1998), *Raum-zeitliche Rekonstruktion von Witterungsanomalien und Naturkatastrophen 1496-1995*, Zürich: Vdf Hochschul-Verlag an der ETH.

Phung, Thuy Phuong (2002), *Ecological Modernisation of Industrial Estates in Viet Nam,* PhD dissertation, Wageningen: Wageningen University.

Ponting, Clive (1991), *A Green History of the World: The Environment and the Collapse of Great Civilizations*, New York: St. Martin's Press.

Princen, Thomas (1997a), "The Shading and Distancing of Commerce: When Internalization is not Enough", *Ecological Economics,* vol. 20, no. 3.

Princen, Thomas (1997b), "Toward a Theory of Restraint", *Population and Environment*, vol. 18, no. 3.

Princen, Thomas (1998), "Consumption, Ecological Constraint, and the Nature of Work", *Annual convention of the International Studies Association*, March 18-21, 1998, http://www.lancs.ac.uk/users/scistud/esf/princht.htm

Princen, Thomas (1999), "Consumption and Environment: Some Conceptual Issues", *Ecological Economics,* vol. 31, no. 3.

Princen, Thomas (2001), "Consumption and its Externalities: Where Economy Meets Ecology", *Global Environmental Politics*, vol. 1, no. 3.

Princen, Thomas; Maniates, Michael; Conca, Ken (eds.) (2002), *Confronting Consumption*, Cambridge, Mass.: MIT Press.

Renn, Ortwin (1991a), "Risk Communication and the Social Amplification of Risk", in: Kasperson, Roger E. and Stallen, Pieter Jan M. (eds.), *Communicating Risks to the Public: International Perspectives,* Dordrecht: Kluwer Academic Publishers.

Renn, Ortwin and Levine, Debra (1991b), "Credibility and Trust in Risk Communication", in: Kasperson, Roger E. and Stallen, Pieter Jan M. (eds.), *Communicating Risks to the Public: International Perspectives,* Dordrecht: Kluwer Academic Publishers.

Renn, Ortwin and Zwick, Michael (eds.) (1997), *Risiko- und Technikakzeptanz*, Berlin: Springer-Verlag.

Renn, Ortwin (2004), Lecture given at the Fifth World Congress of Foodborne Infections and Intoxications, German Federal Institute of Risk Assessment, 7-11 June 2004, Berlin.

Robbins, Richard (1999), *Global Problems and the Culture of Capitalism,* Boston: Allyn & Bacon.

Røpke, Inge (1999), "The Dynamics of Willingness to Consume", *Ecological Economics,* vol. 28, no. 3.

Royal Society of London and the United States National Academy of Sciences (1997), *Towards Sustainable Consumption*, London: The Royal Society.

Sahlins, Marshall (1974), *Stone Age Economics*, London: Tavistock.

Schmidt-Bleek, Friedrich (1999), "The MIPS-Concept: Bridging Ecological, Economic, and Social Dimensions with Sustainability Indicators", Paper for the Euroconference "Quality of Life, Sustainability, Environmental Changes", Stadtschlaining, Austria, 6-9 May 1999.

Schnaiberg, Allan (1980), "The Expansion of Consumption", in: Goodwin, Neva; Ackerman, Frank; Kiron, David (eds.) (1997), *The Consumer Society*, Washington D.C.: Island Press.

Schneider, Hartmut (ed.) (1993), *Environmental Education: An Approach to Sustainable Development*, Paris: OECD.

Schor, Juliet (1991), *The Overworked American: The Unexpected Decline of Leisure*, New York: Basic Books.

Schor, Juliet (1994), "New Analytic Bases for an Economic Critique of Consumer Society", in: Goodwin, Neva; Ackerman, Frank; Kiron, David (eds.) (1997), *The Consumer Society*, Washington D.C.: Island Press.

Schor, Juliet (1995), "Can the North Stop Consumption Growth? Escaping the Cycle of Work and Spend", in: Bhasker, V., Glyn, Andrew (eds.), *The North, the South, and the Environment: Ecological Constraints and the Global Economy*, New York: St. Martin's Press.

Schor, Juliet (1998), *The Overspent American: Upscaling, Downshifting and the New Consumer*, New York: Basic Books.

Schwarz, Jürg and Spreng, Daniel (1996), "Einfache Modellierung der Entwicklung des Energieverbrauchs von Energieanwendungen", *Bulletin SEV/VSE*, no. 22.

Segal, Jerome (1994), "Alternatives to Mass Consumption", in: Goodwin, Neva; Ackerman, Frank; Kiron, David (eds.) (1997), *The Consumer Society*, Washington D.C.: Island Press.
Schipper, Lee (1997), *Indicators of Energy Use and Efficiency*, Paris: OECD/IEA.
Shove, Elizabeth (1997), "Revealing the Invisible: Sociology, Energy, and the Environment", in: Graham, Michael Redclift (ed.), *The International Handbook of Environmental Sociology*, Cheltenham: Elgar.
Shove, Elizabeth; Lutzenhiser, Loren; Guy, Simon; Hackett, Bruce; Wilhite, Harold (1998), "Energy and Social Systems", in: Rayner, Steve and Malone, Elizabeth L. (eds.), *Human Choice and Climate Change*, vol. II, Columbus: Battelle Press.
Spaargaren, Gert (1996), *The Ecological Modernization of Production and Consumption*, PhD dissertation, Wageningen: Wageningen University.
Spaargaren Gert (2000a), "Ecological Modernization Theory and the Changing Discourse on Environment and Modernity", in: Spaargaren, Gert; Mol, Arthur P.J.; Buttel, Frederick H. (eds.), *Environment and Global Modernity*, London: Sage.
Spaargaren, Gert and van Vliet, Bas J.M. (2000b), "Lifestyles, Consumption and the Environment: The Ecological Modernisation of Domestic Consumption", *Environmental Politics*, vol. 9, no. 1.
Spaargaren, Gert (2003), "Sustainable Consumption: A Theoretical and Environmental Policy Perspective", *Journal of Society and Natural Resources*, vol. 16, no. 8.
Spangenberg, Joachim (1995), *Towards Sustainable Europe; The Study*, Wuppertal Institute for Climate, Environment, Energy/Friends of the Earth Europe.
Spreng, Daniel (1988), *Net Energy Analysis and the Energy Requirements of Energy Systems*, New York: Praeger.
Spreng, Daniel (1994), "Energiebewirtschaftung: Strategien für die Zukunft" in: Schregenberger, Johann W. (ed.), *Die Zukunft beginnt im Kopf: Wissenschaft und Technik für die Gesellschaft von morgen*, Zurich: Vdf Hochschulverlag an der ETH.
Spreng, Daniel and Goldblatt, David L. (1998), *Innovations in Energy Efficiency: Theory and Applications in Switzerland and Abroad*, Bern: Swiss Federal Office of Energy.
Spreng, Daniel; Truffer, Bernhard; Wüstenhagen, Rolf (2001), "Perspektiven für die Wasserkraftwerke in der Schweiz: Die Chancen des Ökostrommarktes", Zurich: CEPE, Swiss Federal Institute of Technology.
Stern, Paul C. et. al. (1987), "Managing Scarce Environmental Resources", in: Stokols, Daniel (ed.), *Handbook of Environmental Psychology*, New York: Wiley.
Stern, Paul C. (1992), "What Psychology Knows About Energy Conservation", *American Psychologist*, vol. 47, no. 10.
Stern, Paul C. et. al. (eds.) (1997a), *Environmentally Significant Consumption: Research Directions*, Washington D.C.: National Academy Press.
Stern, Paul C. (1997b), "Toward a Working Definition of Consumption for Environmental Research and Policy", in: Stern, P.C. et. al. (eds.), *Environmentally Significant Consumption: Research Directions*, Washington D.C.: National Academy Press.
Stern, Paul C. (1997c), "Examining the Driving Forces: Introduction", in: Stern, Paul C. et. al. (eds.), *Environmentally Significant Consumption: Research Directions*, Washington D.C.: National Academy Press.
Swearengen, Jack C. and Woodhouse, Edward (2003), "Overconsumption as an Ethical Challenge for Engineering Education," *International Journal of Mechanical Engineering Education*, vol. 31, number 1.
Swiss Federal Statistical Office (2002), http://www.statistik.admin.ch/eindex.htm.
Symposium Sustainable Consumption (19-20 January 1994), Oslo.

Tatum, Jesse (1995), *Energy Possibilities: Rethinking Alternatives and the Choice-Making Process*, Albany: State University of New York.

Uchitelle, Louis (May 3 1998), "Keeping up with the Gateses? Lifestyles of the Rich, Sought by the Middle Class", *New York Times.*

Uiterkamp, A.J.M. Schoot and Noorman, Klaas Jan (1998), "Diagnosing and Evaluating Household Metabolism", in: Noorman, Klaas Jan and Uiterkamp, A.J.M. Schoot (eds.), *Green Households? Domestic Consumers, Environment, & Sustainability,* London: Earthscan.

Unruh, Gregory C. (2000), "Understanding Carbon Lock-In", *Energy Policy,* vol. 28, no. 12.

Uusitalo, L. (1982), "Environmental Impacts of Changes in Consumption Styles", *Journal of Macromarketing*, Fall.

van den Burg, Sander W.K.; Mol, Arthur P.J.; Spaargaren, Gert (2001), "Consumer-Oriented Monitoring and Environmental Reform", Paper presented at the ISA-conference "New Natures, New Cultures, New Technologies", Cambridge University, Jul 5-7 2001.

van der Wal, Jack and Noorman, Klaas Jan (1998), "Analysis of Household Metabolic Flows", in: Noorman, Klaas Jan and Uiterkamp, A.J.M. Schoot (eds.), *Green Households? Domestic Consumers, Environment, & Sustainability,* London: Earthscan.

van Diepen, Albertine M.L. (1998), "Developments in Household Composition in Europe", in: Noorman, Klaas Jan and Uiterkamp, A.J.M. Schoot (eds.), *Green Households? Domestic Consumers, Environment, & Sustainability,* London: Earthscan.

van Vliet, Bas (2002), *Greening the Grid: The Ecological Modernisation of Network-bound Systems,* PhD dissertation, Wageningen: Wageningen University.

Verbruggen, H. and Kuik, O. (1991), "Indicators of sustainable development: an Overview" In: Kuik, O. and Verbruggen, H. (eds.), *In Search of Indicators of Sustainable Development*, Dordrecht: Kluwer Academic Publishers.

Voss, Alfred; Weber, Christoph; Perrels, Adriaan; Faucheux, Sylvie; O'Connor, Martin (1998), "Consumers' Lifestyles and Pollutant Emissions: Summary Final Report", *Cordis Document Library* EV5V-CT94-0373, http://www.npk.gov.pl/cordis/www.cordis.lu/env/src/p-rep6.htm

Vringer, Kees and Blok, Kornelis (1995a), "Consumption and Energy-Requirement: A Time Series for Households in the Netherlands from 1948 to 1992", Utrecht: Utrecht University.

Vringer, Kees and Blok, Kornelis (1995b), "The Direct and Indirect Energy Requirements of Households in the Netherlands", *Energy Policy*, vol. 23, no. 10.

Wachtel, Paul (1983), *The Poverty of Affluence: A Psychological Portrait of the American Way of Life*, London: Free Press.

Wackernagel, Mathis and Rees, William (1997), *Unser Ökologischer Fussabdruck*, Basel: Birkhäuser.

Weber, Christoph and Perrels, Adriaan (2000), "Modelling Lifestyle Effects on Energy Demand and Related Emissions", *Energy Policy*, vol. 28, no. 8.

Whiston, Thomas (1990), "The Global Environment: Technical Fix or Radical Change?", Social Audit Paper 2, Brighton: Education Network for Environment and Development.

Whiston Thomas (2000), "Environmental Consciousness and the Challenges and Barriers to Communicating Environmental Knowledge/Indicators", Paper for the Euroconference "Quality of Life, Sustainability, Environmental Changes", Stadtschlaining, Austria, 6-9 May 1999.

Wilhite, Harold et. al. (1996), "A Cross-Cultural Analysis of Household Energy use Behaviour in Japan and Norway", *Energy Policy*, vol. 24, no. 9.

Wilhite, Harold and Shove, Elizabeth (1998), "Understanding Energy Consumption: Beyond Technology and Economics", *ACEEE 1998 Summer Study Proceedings*, Washington D.C.: American Council for an Energy-Efficient Economy.

Wilhite Harold; Shove, Elizabeth; Lutzenhiser, Loren; Kempton, Willett (2000), "The Legacy of Twenty Years of Energy Demand Management: We Know More about Individual Behaviour but Next to Nothing about Demand", in: Jochem, Eberhard; Satheye, Jayant; Bouille, Daniel (eds.), *Society, Behaviour, and Climate Change Mitigation,* Dordrecht: Kluwer Academic Publishers.

Wilk, Richard (1999), "Towards A Useful Multigenic Theory of Consumption", *Energy Efficiency and CO_2 Reduction: The Dimensions of the Social Challenge: Proceedings of the 1999 ECEEE Summer Study*, Paris: European Council for an Energy-Efficient Economy.

Wilting, Harry C. and Biesiot, Wouter (1998), "Household Energy Requirements", in: Noorman, Klaas Jan and Uiterkamp, A.J.M. Schoot (eds.), *Green Households? Domestic Consumers, Environment, & Sustainability,* London: Earthscan.

Winner, Langdon (1986), *The Whale and the Reactor: The Search for Limits in an Age of High Technology*, Chicago: University of Chicago Press.

World Resources Institute (1995), "Natural Resource Consumption", in: Goodwin, Neva; Ackerman, Frank; Kiron, David (eds.) (1997), *The Consumer Society*, Washington D.C.: Island Press.

Wynne, Derek (1998), *Leisure, Lifestyle, and the New Middle Class: A Case Study*, London: Routledge.

Yankelovich, Daniel (2004), "Speaking Truth to Power", Speech to the AAAS Forum on Science & Technology, April 22 2004, Washington D.C. http://www.aaas.org/spp/rd/yankel404.pdf

Yoe, Charles (2003) Society for Risk Analysis' "Introduction to Risk Analysis," on-line course, November 2003.

Index

Alliance for Global Sustainability Series

1. F. Moavenzadeh, K. Hanaki and P. Baccini (eds.): *Future Cities: Dynamics and Sustainability*. 2002 ISBN 1-4020-0540-7
2. L. Molina (ed.): *Air Quality in the Mexico Megacity*. An Integrated Assessment 2002 ISBN 1-4020-0452-4
3. W. Wimmer and R. Züst: *ECODESIGN Pilot*. Product-Investigation, Learning - and Optimization-Tool for Sustainable Product Development with CD-ROM. 2003 ISBN 1-4020-0965-8
4. B. Eliasson and Y. Lee (eds.): *Integrated Assessment of Sustainable Energy Systems in China*. The China Technology Program. A Framework for Decision Support in the Electric Sector of Shandong Province . 2003 ISBN 1-4020-1198-9
5. M. Keiner; C. Zegras, W.A. Schmid and D. Salmerón (eds.): *From Understanding to Action*. Sustainable Urban Development in Medium-Sized Cities in Africa and Latin America . 2004 ISBN 1-4020-2879-2
6. W. Wimmer, R. Züst and K. M. Lee : *ECODESIGN Implementation*. A Systematic Guidance on Integrating Environmental Considerations into Product Development. 2004 ISBN 1-4020-3070-3
7. D.L. Goldblatt: *Sustainable Energy Consumption and Society*. Personal, Technological, or Social Change?. 2004 ISBN 1-4020-3086-X

29, 38, 40-41, 47, 50, 51, 53, 54, 58
66, 67, 78-79